Tales of a Journeyman Naturalist

Tales of a Journeyman Naturalist

Terry C. Maxwell

Copyright © 2016 Terry C. Maxwell
All Rights Reserved

ISBN: 978-1-942956-35-8
Library of Congress Control Number: 2016957129

Manufactured in the United States

Lamar University Literary Press
Beaumont, Texas

Acknowledgments

How is it possible to acknowledge each person in a lifetime of supporters, enablers, constructive critics, and companions without whom my experiences detailed here would never have happened or perhaps not survived? I cannot name them all but there are some that I must recognize.

I cherish my parents, George Maxwell and Echo (Maxwell) Roeder, for giving me my head at an early age. They provided the setting and freedom in which I learned to value chance-taking over comfort or security.

My mentors provided the opportunities and encouragement that guided my development as a naturalist. Most notable in that long list are: Frances Williams, Keith Arnold, Jim Dixon, William Davis, David Schmidly, Richard Baldauf, Gordon Creel, and Chester Rowell. Following formal academic training, professional companions continued to stimulate my skill development. Among my many such naturalist companions, I note here Loren Ammerman, Bonnie Amos, Royce Ballinger, Mike Dixon, Robert Dowler, Mark Engstrom, Alvin Flury, Edie Marsh-Matthews, Ned Strenth, Wilmot Thornton, and Clarence Wiedenfeld.

With genuine sincerity, I thank the U.S. Air Force for training me in the Spanish language. Although I was not their most accomplished student, I learned enough of that language to make possible much of which I write in these pages.

For the field expedition experiences I relate here, there were always those in charge who tolerated and perhaps even valued my participation. In that group, I am grateful to: Richard LaVal (Central America, 1967), Steven Cardiff (Peru, 1983), Angelo Capparella (Peru, 1984), Mark Robbins (Ecuador, 1988, 1989, 1990, 1991), Mark Engstrom (Guyana, 1994), and Robert Dowler (Galapagos Islands, 2000).

Portions of this book were thankfully reviewed by Loren Ammerman, Robert Dowler, and Ann Maxwell. Their comments were always important, but it is true that any lapses in memory or construction of this work are all my own. Kiah Rhea edited the manuscript and contributed to its readability. Marcia Revelez was tolerant and kind when assisting me with my profound computer ignorance, a serious malady in this age.

For Ann Maxwell

Recent Nonfiction from Lamar University Press

Jean Andrews, *High Tides, Low Tides*
Robert Murray Davis, *Levels of Incompetence: An Academic Life*
Ted L. Estess, *Fishing Spirit Lake*
Dominique Inge, *A Garden on the Brazos*
Jim McJunkin, *Deep Sleep*
Jeanetta Calhoun Mish, *Oklahomeland*
Jim Sanderson, *Sanderson's Fiction Writing Manual*
Steven Schroeder, *What's Love Got to Do With It? A city out of thin air*

For Information on these and other books, go to
www.lamar.edu/literarypress

Preface

This book is a narrative of some events in my life as a naturalist. And although I believe it has not been that out of the ordinary among members of the profession, I have been told that it might be entertaining to others because in pursuit of our mission, naturalists often place themselves into interesting and difficult circumstances that can, in retrospect, make for a good read.

I think in the broad perspective there is nothing unusual about my interests. More or less, nearly everyone is a naturalist. You may be dying to know the kind of snake that scared you last week. You wonder what plant sported that red flower you saw down by the gate this morning. It's just that those of us who call ourselves naturalists carry our love of nature to, what you might well call, extreme. We are absorbed by it. We dream about it. We pick vacation spots based on our obsession with it. We put ourselves in harm's way in pursuit of it. We may even strive to make it the subject of our professional life.

Every naturalist's approach to his or her passion is different. Some, for example, are content to identify the birds or flowering plants in the neighborhood. Others want to know all the birds that are found in their home state. Still, others want to learn something about the behavior, diet, life cycle, or more about some particular species and maybe even get paid to do that. There are many ways to be a naturalist. After a false start or two, I decided to become a university biologist. In my own case, professional life came to be dominated by classroom teaching. Much of my research, particularly abroad, tended to follow a "journeyman" approach. Together with investigation of natural history close to home in West-central Texas, this approach was a distinctive way of life that accumulated memories I share here with you.

Since medieval time, a journeyman has been one skilled in a trade or craft but who works for another who is in control of the greater effort. For the better part of four decades, I was a glad journeyman in the world of vertebrate faunal investigation. My students and I studied woodpeckers, orioles, wrens, and communities of birds in west-central and more broadly in Texas. At other times, I taught as a universityfaculty member from September to May then often flew or drove to distant places throughout the western hemisphere for portions of many summers. There I typically observed birds or collected specimens of birds and mammals. I worked, usually, for doctoral graduate students or faculty scientists at Angelo State University, Texas A&M University, Louisiana State University, or Academy of Natural Sciences in Philadelphia. They shouldered the load of analysis and publication. For my interests and limited skills, it was the ideal passage of a naturalist's life beyond the classroom.

     My youth in the dry brushlands of West-central Texas had, I think, a formative effect on what I became although I don't understand, to any mature degree, the arguments about nature versus nurture in the shaping of our lives. Was I born genetically pre-set for a life absorbed with wild animals, or did my early experiences tug me that way? Might my innate abilities have been better suited for a career as a lawyer or physician rather than a teaching naturalist? Almost surely I would have earned more income with either of those careers, but I took another path, and not once looked back, until now.

CONTENTS

ix	Preface
13	Origin
16	Duchess
18	Bird Nerd
22	Aggieland
24	La Libertad to Danlí
34	Uniformed Interlude
37	Eagle Trouble
43	Quebrada Vainilla
55	North Slope
65	Esmeraldas to Zamora Chinchipe
80	Rio Essequibo
93	Islas Encantadas
109	Bits and Pieces
121	Civilized Journeys of Exploration
128	Good Times

List of Figures

28	1. great fruit-eating bat warming and drying on my head
30	2. diversity of bats in the American tropics includes common vampire bat and wrinkle-faced bat
39	3. golden eagle
47	4. Shaded floor of the tropical rainforest
52	5. horned screamer
53	6. Amazonian umbrellabird
57	7. frozen Peter's Lake, Arctic National Wildlife Refuge in Alaska
57	8. camp on the North Slope near the Aichilik River
60	9. nesting birds, North Slope tundra
61	10. Willow ptarmigan
63	11. hitting the antenna on Barter Island in the Beaufort Sea
75	12. size variation within the woodpecker family
77	13. toucans and their close relatives
79	14. endemic tropical American bird families
83	15. Guyana team awaiting departure
87	16. white-plumed antbird feeds on arthropods

87	17. screaming piha
89	18. amphisbaenian or worm lizard
91	19. royal flycatcher
94	20. Galapagos mockingbird
97	21. large ground finch of the Galapagos Islands
101	22. flightless cormorant of the Galapagos Islands
102	23. huge Galapagos tortoise
113	24. elf owl, smallest of the world's owls
117	25. into the maw of the Devil's Sinkhole
119	26. zone-tailed hawk and common black-hawk

Origin

I was born in San Angelo, Texas in 1945 to a 22-year-old mother from rural hardscrabble Coleman County and a father recently returned from the war skies above Nazi-occupied Europe. In my memory of the childhood they gave me, most of our life together was happy, even though we were, by many standards, of moderate to low income.

My dad, George Maxwell, dropped out of high school during the Great Depression to help support his parents and siblings. Quitting school was his defining moment, one that I believe drove him to compensate by competing in many aspects of his life.

He married my mother, Echo Box, in 1942. Shortly after my birth, we lived for a brief period in coastal Mississippi but returned to San Angelo and moved into a three room cinder block house built by dad and my paternal grandfather. Our home had a kitchen, bathroom, and one big living room in which we also slept.

My sister, Carol Lynn, came along in 1948. Dad supported us all as a mail carrier in the U.S. Post Office. Later, he found the money to build a tin chicken barn that would hold a thousand or more birds, and we entered the commercial chicken business for supplemental income. Maybe that was it. Perhaps my chores of watering and feeding 1000 chickens inspired me to a life with wild birds, but thinking back on how much I despised those fowl, probably not.

My most vivid memories of that chicken barn are of a railroad train, a tornado, a monkey, and my mother's courage. Among my morning chores was staking out the milk cow where grazing was best. Sometimes that was on the nearby railroad track right of way, which was okay since the train only came along around midnight. My main job in the late afternoon was to get the cow back to her pen, a task I forgot one day. We all sat bolt upright in our beds that night as a stopped Santa Fe train blew its ungodly horn at a bewildered cow staring into the train's headlight. The bigger issue was the several hundred chickens that suffocated. In their fright of the train's horn they stacked up on the walls of that barn.

My father was not amused. We tried to save what we could of his profit margin by scalding the feathers off hundreds of chickens and then selling the fryer carcasses around town.

Another memory of that beloved cow was my mother's braving of a severe hail storm to get the cow to cover. With my sister and me anxiously watching out a window, mom covered her head with a sheet of tin, ran to the pen and pulled the cow up onto the covered front porch of the house. Like I said, my mother never lacked for courage.

On May 11, 1953 mom was in the barn when she heard a roaring sound. Curious, she hurried outside and saw an approaching tornadic storm. Grabbing my toddler sister from the house, she ran to a railroad culvert where the two suffered in dangerously high water until it passed. Although the tornado jumped over our property, thirteen were killed and over 150 injured in San Angelo that terrible day.

At another time, something possessed her to grab hold of a monkey she found cowering in the rafters of that barn. Before she returned it to the pet store from which it had escaped, it bit her, savagely. But mom held on. I suspect that monkey was the only one ever captured on the southern plains of Texas.

I grew from tricycle motor to early teen years during the great drought of the 1950s. The first rain storm I remember, in 1957, I was 11. It was not until well into my second decade that I learned dust storms did not have to be the norm in spring. In one of the severe haboob storms, the post office shut down for the day (contrary to the legend of postal service never stopping). My father drove home and bumped the car into the front porch, which he could not see for the blowing wall of dirt. Wet cloths along window seals hardly helped; dust got into our bedding, in our kitchen cabinets, everywhere. Never since have I experienced such dust storms.

My grade school was north of San Angelo in the rural community of Grape Creek. There were eight grades in our three-room building. First through fourth grades were in one room, each grade in its own line of desks. Fifth and sixth were in another room, and the big kids, seventh and eighth, in the remaining room. My little sister and I rode the school bus for an hour in the morning and an hour in the afternoon every school day.

My fondest memories of that school were of the county library bookmobile and fieldtrips to the Coca Cola bottling plant (a magical place). The bookmobile introduced me to the world. We were too poor to afford a subscription to National Geographic magazine, but checking out arm-loads of books from the bookmobile was free.

When I was about nine, dad built a new wooden house close to our old one so we could all have our own bedrooms. The next three years seemed to have been ones of contentment, until my parents divorced. My mother remarried and bore my brother David when I was seventeen. Dad later also remarried—unsuccessfully again.

One way for children to deal with the strangeness of divorce—the one I followed—was to concentrate on their own world. For me that meant plenty of interesting animals to examine: mockingbirds, whiptail lizards, toads, longhorn beetles. Although confused and sad, I don't think I was ever seriously angry with my parents over divorcing; I was too distracted in my private world.

A defining feature of my life has been drawing. As a child, my grandfather Maxwell considered it his duty to teach me to draw people's portraits. It's the only contact with him that I remember. In retrospect, I wasn't very good at it, and I still cannot draw a face that looks like the actual owner. Three of his children were artists. My aunts Polly and Betty Jane had some formal training and were quite good. Dad was good enough to teach landscape painting in his retirement years.

All of my artistic efforts—at least those with any success—were in drawings of animals, especially of birds. Pen-and-ink and chalk were my tools of choice. I struggled with color, and decided long ago to not call myself an artist but rather an illustrator. Family and friends insist otherwise, but then they are, after all, family and friends.

Duchess

I grew up with a dog, a border collie named Duchess which I loved beyond all reason. She was my companion from dawn to dark. We enjoyed most of the same things: lizards, toads, jackrabbits and hamburgers, although I was expressly forbidden to give her any of mine, and I never could explain how she always managed a bite. We were both wary of oatmeal, asparagus, rattlesnakes, mesquite thorns, and bumble bees. I never saw her pay the slightest attention to birds, an attitude that was among her few failings.

Mom had a small flowerbed, valued by armadillos and toads. The armadillos were looking for grubs. The toads were looking for moist loose soil to dig down into for the winter, a place to hibernate. Each spring they would emerge from that flowerbed with the first night of soaking rain at just the right ambient temperature.

Duchess was on deck for most of those nocturnal toad appearences. Everyone in the house would awaken to the dog excitedly barking her head off. Dad demanded I find a way to shut her up. So often I found myself at two in the morning in my underwear out on a wet lawn, gathering toads.

Duchess and I chased whiptail lizards at every opportunity. I did not know then that they specifically were Texas spotted whiptails but I would come to that understanding with the formal training that awaited me in college. On these childhood hunts it was enough for us to chase them under bushes and dive for the capture. Neither of us ever caught one. Roadrunners are masters at catching them, but they have patience, something Duchess and I lacked.

Mrs. Emmet lived several hundred yards down our caliche road, and she often had fresh baked cookies. Mom had parental responsibilities, so it wasn't hard to find an excuse to be mad at her and run away to Mrs. Emmet's house. Of course, Duchess ran away with me;

there might, after all, be lizards on the way. My plan for the getaway involved creeping low behind bushes. Little did I know that the dog's high wagging tail meant we were not hidden at all. I found out in later years that mom would call Mrs. Emmet and tell her "he's on his way to your house," to which Mrs. Emmet would respond "I'll watch him till you can come get him." Watching me involved cookies. I think that I usually got a spanking over that when dad got home, but the pleasure of fresh cookies necessarily had a cost. I was quite young when I learned that life often involves a matter of tradeoffs.

The day came when San Angelo extended its city limits, and our property was included. It was of no obvious consequence to me—with one wrenching exception. My devoted companion's status as a free creature came to an end. A city ordinance required that Duchess wear a collar with proof of rabies vaccination. I regarded the binding of Duchess with a wretched collar was an attack on both of our rights. I regarded it as a form of unjust imprisonment for Duchess. Of course I got over it, but she did so first, for she was better at not holding grudges.

There was innocence to the time and location of my childhood. Out in the country in mid-20th century, the threats to a child could be real but seem so less disturbing than those faced by an urban child today. Apparently, my parents accepted that I understood the danger of rattlesnakes and falling into a prickly-pear cactus patch. I remember having freedom to roam in the brush, freedom to explore the huge dam behind our house, freedom to be a confident me.

I think developing such confidence put me in good stead later in life from Alaska to Peru.

Bird Nerd

Nerds are not supposed to fit in with normal kids. They can be inordinately intellectual or social misfits. In my youth, a nerd was often conspicuous by the slide rule in a scabbard on his belt, for those were pre-calculator days. Nerds always turned homework in on time, always. They also spend plenty of time on obscure or "non–mainstream" activities, surely a teenage sin all periods of time. Want to guess where bird watching fit in to the high school mainstream social structure?

I don't remember when I became interested in birds, other than that I was young. I do, however, remember a book I was given for Christmas: *Birds of America,* edited by T. Gilbert Pearson. I still have it and it must weigh 10 pounds with its 296 pages and 106 full color plates by Louis Agassiz Fuertes. Fuertes remains, in my estimation, the greatest American bird illustrator. His raptors are the finest renditions ever done of such birds. I studied all 106 plates and read every word many times.

The family's story was that my father bought the book for me, and my mother thought he had lost his mind. It certainly was not a child's book. But then my interest in birds had transcended a child's approach. I couldn't guess whether his choice of a gift was just dumb luck or if he saw something in me that eluded my mother.

My high school science fair project was entitled "Ornithological Investigations in Tom Green County." On poster board, I listed over 150 species of birds along with their seasonal status and primary habitat in the county. I had been collecting the information for years, since late grade school. I saw every one of those birds and filled up notebooks of detailed observations. Most people who saw the presentation were incredulous. I tied for first place in the city competition but came in second to a good physics experiment at regional. I overheard one contestant's father commenting to his son that he was pretty sure that things with exotic names like loon, avocet, and godwit didn't really occur here. They didn't understand what a dedicated bird nerd could discover in his own back yard.

Finding someone to learn from or in any way interact with about birds in San Angelo in the 1950s and 1960s was a challenge for a teenager. But then somehow I met three women of the Midland Naturalists. My first birding trip with Anne LeSassier, Ola Dublin Haynes, and Frances Williams was an education in itself. LeSassier and Williams were in their 40s, and Haynes was older. They were regional cooperators with Edgar Kincaid, who then was updating information on the bird distributions in all of Texas' 254 counties in preparation for publication of H. C. Oberholser's monumental *The Bird Life of Texas*. The Midnats spent most weekends chronicling the birds of the Midland-Odessa region, county by county on those bleak, windy southern plains.

It was important for me to bird with them and to learn some of what they knew. They taught me to send observations to *Audubon Field Notes*, a journal devoted to birding records. Frances was editor of the southern plains reporting region in that journal. She also was coauthor of the species account of Cassin's Sparrow in Bent's legendary *Life Histories of North American Birds*. Frances Williams was my first mentor and inspiration in ornithology.

To emphasize an earlier point, try explaining to the High School social net a teenage boy spending his weekends looking at birds (not killing, just looking) with middle-aged women who were fellow bird nerds. I wasn't old enough to ignore peer pressure, so I kept such activities to myself.

I doubt that I ever met a birder who seriously played high school sports. There must be some, but I never met them. I actually tried after my father lamented that an athletic scholarship was probably my only chance at affording college. My single appearance in a San Angelo Central High School football game lasted about two minutes and resulted mostly in my father's embarrassment. The coach, Emory Bellard, who went on to coach Texas A&M and Mississippi State Universities, commented that my actions on the field resembled a chicken with its head cut off. Those blamed chickens again. Clearly, I was going to have to find other funding for college.

For bird nerds, after school and weekends were for birding, when not working. What drove me most was listing: adding birds to a list of species I actually saw, my personal life list. It's sort of like coin or stamp collecting only you collect sightings of species. But instead of surfing through sales catalogs that offered your desired coins or stamps, you had to lace up your boots and hike into the brush. You must see the bird in habitat to add it to your life list. For many of us, birding actually was a competitive sport without blocking or tackling but with serious rules.

Memory of my high school days is now pretty much a blur of school work, shelving books after school at the county library, and birding. I did discover Big Bend National Park. It was just short of 300 miles from my home in San Angelo to the park's Chisos Mountains basin. I lost count of the number of my nocturnal round trips down there on Friday after work and back home on Sunday evening. My stepfather had bought me a Ford Fairlane 500, and it would hum at ridiculous life-threatening speed between my home and the park.

In Big Bend you could find Colima Warbler and Mexican Jay, birds of exotic Mexico. Elf owls, black-tailed gnatcatchers, varied buntings, crissal thrashers, and more were there among Spanish dagger, creosote bushes, and alligator juniper. The desert has a unique allure that naturalists find hard to explain. Certainly part of it is the surprising diversity of life forms in an environmentally challenging climate. There's also a crisp cleanness to the morning desert air.

The Texas birding fraternity did not network by computer in those days. No one did. Personal computers did not really become available until 1977, nine years after my college graduation, and did not become affordable home devices until the early 1980s, well after I had begun my professorial career. Even then, it was not until the mid-1990s that the internet pretty much came to be what we know today.

Before the internet, birders and other naturalists came to know each other through clubs, newsletters, Audubon Field Notes, and Audubon Christmas Bird Counts. We didn't have the slightest notion that we were deprived without the gratification of instant contact. If you wanted, you could befriend a wide array of Texas birders.

As my awareness grew about who else was out there and what they did, I came to know people with all sorts of professions—academics, scientists, lawyers, ranchers, nature tour guides, and more. I decided that my place in the larger scheme of things would be to work for a government agency while wearing a Smoky Bear hat—maybe as a park naturalist. So I graduated high school, and armed with a small scholarship from the West Texas A&M Club, headed for the Wildlife Science program at Texas A&M University. It was a career-defining decision.

Aggieland

I will always remember my first day on the Texas A&M University campus. It was alarming. As with all incoming freshmen in 1964, I was in the Corp of Cadets, uniformed and all. I couldn't turn a corner that first afternoon without being yelled at by some upper classman in Squadron 6. Never before had I concerned myself with shirt tails and collars and shoe polish, but there now seemed to be surly sophomores in every hallway determined to see that I came to care.

Then I tried to sleep. To a kid from the dry southern plains, College Station in August is a steam cooker. Sleep was fitful, and I got up most mornings with my sheets soaked in sweat. At that point I would never have dreamed of loving the school and its natural surroundings, but I soon did.

Escaping cadet sophomores became essential for my sanity, so I employed my tried and true approach—immersion in nature. A&M is situated in the Southern Post Oak Savanna, about thirty miles from piney woods and only 160 miles from Louisiana—not far in my estimation. In the 1960s back home, people would drive 200 miles to Austin or San Antonio just to go shopping. Ecologically, it was a far cry from the mesquite brushlands of San Angelo and being at A&M provided my first extended experiences in a true woodland biome.

I soon discovered the Navasota River bottom. Living there were salamanders, tree frogs, three-toed box turtles, eastern hognose snakes, and copperheads. Duchess would have loved it. The great pileated woodpeckers dwarfed any woodpecker back home. Prothonotary warblers were so golden they glowed. It was rare on a Saturday morning to find me anywhere other than in the river bottom.

Another place you could often find me was in the wildlife laboratories. The wildlife department in those days was housed in the old agricultural engineering building. Why they let us camp in there I'll never understand. My favorite labs reeked of formaldehyde and Doc Davis's cigar stubs. I remember when the big indigo snake escaped and terrorized the secretaries, though none of them ever saw it.

When I was a freshman in 1964, the department head was William B. Davis, a legend in Texas mammalogy. Doc Davis authored one of my natural history bibles, *Mammals of Texas,* published by the Texas Parks and Wildlife Department. Doc was a taxonomist, largely interested in bats. What a taxonomist in general does is explore the world for evidence of what species occur where, describe new species, and place them into the right slot or category within the evolutionary tree of life. It did not take me long in that heady environment to discard my dream of the Smokey Bear hat for something akin to the life of Doc Davis.

Academically, my freshman year was a disaster. I don't recall all my less than stellar grades, but I do remember the "D" I didn't deserve in chemistry. It was an out-and-out gift for what was a thudding F average. My academic advisor, Jack Inglis, stunned me after a disappointing semester with advice to quit and join the navy. I didn't take him up on it, and by the middle of my sophomore year I had learned to be a successful college student.

My interest in birds remained strong, so I naturally gravitated to the department's scientific bird collection. The ornithologist before my arrival had gone on to another school and a new one had not yet been hired, so a graduate student, Al Bjelland, was care-taking the collection. He let me help, and I became deeply committed. From then until the end of my career, scientific bird collecting dominated the research component of my life.

A scientific collection is an accumulation of specimens: stuffed and dried skins, skeletons, and spirit- (alcohol) preserved animals. In effect, they are stored examples of the variety of life forms. The information associated with each specimen must include where and when it was collected, its sex, and who collected it. Additional information, like measurements, stomach contents, habitat, and so forth can be included. Collections are the primary storehouse of information about the organism's anatomy, seasonal activities, diet, and for many groups, its distribution.

At the Academy of Natural Sciences in Philadelphia, I have held specimens of extinct passenger pigeon, Carolina parakeet, and ivory-billed woodpecker. In Paris, France, I stood before the skeleton of an extinct and little known Steller's sea cow from the Bering Sea, and in Tring, England, I was allowed to inspect Galapagos finches prepared by Charles Darwin.

I work today in a building that houses scientific skin specimens of *Nesorysomys* rodents from the Galapagos Islands, thought extinct for more than a century until discovered by my friend and colleague, Robert Dowler. It's an aspect of natural history study that is satisfying and amenable to my few talents. My abilities in specimen preparation were soon to send me deep into Central and South America.

Within a year, Texas A&M's new ornithologist, Keith Arnold, arrived, originally from Michigan and most recently from Louisiana State University. He took up the helm, and I flourished under his direction, although the first specimen preparation assignment he gave me was a horrifically smelly, putrefied gannet found dead on the Texas coast. I skeletonized it, a process that first required removal of thoroughly green decomposing viscera. Good grief, I asked during that process, what had I gotten into?

I took advantage of every opportunity to go afield with those who knew how to study animal life. My ichthyology class seined fishes from the headwaters of the Trinity River north of Dallas. My herpetology class canoed into swamps and, by hand, grabbed large, testy water snakes. I assisted a graduate student in a survey of mammals in remote Capote Canyon whose creek drained into the Rio Grande upstream of Presidio.

It was all the stuff of my dreams, opportunities in which I established my usefulness, professionalism, and drive to do this work. Apparently, my efforts were noticed.

La Libertad to Danlí

My professional life changed forever in spring 1967 when as a college junior I was chosen to be the assistant on an expedition to tropical El Salvador and Honduras. Doc Davis had been awarded a National Science Foundation grant to study bats of the Central American core. That same spring, Dilford Carter, a Wildlife Department mammalogist, and Richard LaVal, a graduate student, traveled with others to the region. Richard was to lead the follow-up summer trip, and I was to be his sole assistant.

I had never before been to Latin America, and did not have even a rudimentary understanding of Spanish. But I could prepare specimens, I had by then considerable field experience, and I desperately wanted to make the trip. I assured my parents that it was a piece of cake and not to worry.

Richard and I left College Station in the third week of June in an A&M suburban loaded with camping and collecting equipment, which included firearms. We stopped the first night in the Lower Rio Grande Valley of Texas. Something I ate there gave me the turista—we hadn't even made it to Mexico, and for at least the next two days of travel I was afflicted with diarrhea and vomiting. You can imagine me on the side of a busy Mexican highway with my pants down—there was no time to stop to let me recover. It was an inauspicious beginning. The remainder of our drive through Mexico proved uneventful. Below Mexico City we picked up the Pan American Highway, which we would follow, dodging gaping potholes, all the way into El Salvador.

The absence of memorable events ended at the Guatemalan border. Guatemala was in its usual political turmoil. A civil war lasted from 1960 to 1996. The year previous to our trip, American Special Forces participated in a Guatemalan counterinsurgency campaign that purportedly killed over 8,000 people. In 1968, the year following our expedition, the U.S. Ambassador was assassinated. La Mano Blanca (the White Hand), a paramilitary right wing death squad, was actively killing perceived leftists, including a lot of professors, students, and indigenous Maya. So here were Richard and I, students both, with weapons on board, driving merrily through all this turmoil on our way to El Salvador. Actually, that's an exaggeration, we were not all that merry, and it was decidedly not a piece of cake. The previous spring Richard and Dilford were thrown temporarily into jail because of their shotguns, legally permitted though they were.

We crossed the border from Mexico into Guatemala in the Sierra de Cuchumatanas. At the border security post, we parked the suburban, got out, and were met by guards with scary-looking weapons and who were warily looking up to the forested slopes above the border post. It came home to me that I was no longer, geographically or politically, in Central Texas. Satisfied with their inspection of our vehicle and permits, we were allowed to enter their country.

The drive to Guatemala City passed through legendary Huehuetenango, originally an ancient Mayan settlement in the mountains. I remember seeing men with huge barrel chests—indicative of high altitude laborers—carrying on their backs massive firewood bundles that reached three or four feet above their heads. Our hotel in Guatemala City was noteworthy for the large magnificent painting of a resplendent quetzal, the national bird of Guatemala, and for the bullet gouges in the front façade of the building.

We made it without mishap out of Guatemala and on to the Pacific coast of El Salvador, our first destination. Our base of operations there was a small pensión (motel) in La Libertad, right on the coast. It had a beautiful black sand beach. One late afternoon with a Pilsner national beer of El Salvador in hand, I watched what I thought were spinner dolphins leap out of the Pacific, twisting in the air. Our rooms in this sumptuous emporium were of a structure that allowed fruit bats to enter and roost above our beds at night where they happily ate fruit and left suspicious substances on our sheets.

I should explain what one does on a collecting trip for bats. Night-flying bats are readily captured in mist nets. Such nets are typically made of nylon thread, are thirty feet long and seven feet deep. The net is strung between two poles and looks a bit like badminton netting. Bats fly into it and become entangled. Although these flying mammals have very good sonar and can detect the net, they still often get caught, presumably because the net is positioned in a location familiar to them and where no net existed before.

The big decisions in such efforts are where to set the nets and how many nets to set for a night's effort. It's also important to get bats out of nets as quickly as possible to prevent excessive entanglement and gaping holes from the bats chewing the net webbing. In tropical forests, the number of nets one attempts to use on any night is critical. On more than one occasion in El Salvador and Honduras I put up as many as twenty nets in a shade tree coffee finca (farm), and paid the price. Taking bats out of nets without being bitten is laborious for a novice (clearly describing me that summer). So twenty-net nights were long sweaty running bouts, net to net, in near panic over how many bats were already hitting nets I had just cleared, not even considering the nets ahead that I hadn't gotten to yet.

Many tropical bats are fruit-eaters. Natural selection often favors those bats that can consume fruit early in their development—and therefore get the fruit before others do. Eating such hard, green fruit requires stout teeth and powerful biting jaw muscles. Taking most bats safely (for the bat and you) out of a net can be a challenge until you've developed your skills at avoiding those teeth.

Sturnira is a genus of small fruit-eating bats that I came to despise. They're actually attractive, often bright orange in color. But their snarly dispositions and agility at getting my finger skin in their mouths distracted from whatever appeal they otherwise had. And then there are the big *Artibeus* fruit-eaters. The Jamaican fruit-eating bat and its larger relative, the great fruit-eating bat, were common at sites we worked in both countries. They're both determined biters.

One night I was checking nets set over a small stream, where I found a female great fruit-eating bat that had been caught in the lowest shelf of the net, resulting in a dunking in the stream. She was wet, shivering, and very pregnant. I momentarily lost my scientific composure and put her into a warm shirt pocket. After she recovered, I was busy removing other bats from nets when she crawled out of the pocket, in the usual bat posture of upside down. She hitched her way to my shoulder, then up my neck, and ended up on top of my head with wings draped over my ears. For a while, I wore a bat bonnet until her muscle warmth was restored and my expectant mother bat flew safely away.

Figure 1. great fruit-eating bat warming and drying on my head

On another occasion, I was seated on the ground at the end of a line of nets, waiting to check them again. My headlamp was off. I heard a scratching sound on the forest floor nearby, as if a small animal was approaching through the leaf litter. When it sounded close enough, I switched on my light and startled a common vampire bat. It quickly flew away. Later that evening, we played the game again, and then again. I don't remember our game count, but the bat was persistent. The species is common in Central America where it feeds on mammal blood. They typically approach their prey on the ground where the bats are surprisingly agile runners. Taking one out of a mist net is tricky; they have a pair of upper incisors that are sharp as razor blades.

We collected for two weeks in El Salvador. All of our sites were within twenty miles of La Libertad. An interesting find was a hairy-legged vampire bat. There are only three species of vampire bats, and this is the only one ever found in the US and then only once, in 1967, the very same year of my Central American expedition. It feeds mostly on bird blood. Otherwise, we made a useful but unspectacular collection of bat species, so we headed into adjacent Honduras.

Our first collecting locale was near La Esperanza, Department of Intibucá. At about 5,500 feet elevation, it is the highest city in the country. We netted for twelve nights, then one night each near Marcala and Jesus de Otoro, both nearby highland sites.

All naturalists have memorable moments in their lives when they first see some rare or spectacular animal or plant. I had one of those moments in a peach orchard in La Esperanza where I stared into the mug of a wrinkle-faced bat. As you can see from my drawing, the hairless face is thrown into grooves and various naked protuberances. It's a fruit-eater that pushes its flat face into ripe fruit, macerates the fruit with its teeth, and swallows mostly juice. It has a ruff of loose throat skin that it pulls up over most of the naked face when roosting.

A patch of thin, almost transparent skin on that ruff is then positioned to form a breathing channel. It's one of the most amazing beasts I've ever witnessed.

Figure 2. diversity of bats in the American tropics includes common vampire bat (upper left) and wrinkle-faced bat (lower left)

We regularly ate the breast meat from *Artibeus* fruit bats. That probably sounds distasteful (or even abhorrent) to many readers, but there's really nothing untoward about it. It was the only flesh that came into our camp, and although gamey in flavor, was acceptable in stew.

Our major locale of operations in Honduras was near Comayagua, about 50 miles from the Honduran capital of Tegucigalpa. There we hired a young man, Eduardo. His job was to help in all ways, and he was a great aide. We netted for 19 nights around Comayagua, working from our field tent. The collecting was good, but it was in Comayagua that I had my most memorable brush with civilian law in any country.

We heard that bats flew out of an old, abandoned church, so we asked the local priest for permission to investigate. He was agreeable and loaned us a key. We checked it out and found a huge colony of gray long-tongued bats roosting behind the altar.

Then the police found us.

Apparently, the residents of that neighborhood were terribly offended. They thought we were desecrating a holy site. We were taken to the police station where we were kept until our story was confirmed by the priest. Although we were treated well, it's still unnerving to be escorted through public streets by an armed police posse.

We drove into Tegucigalpa for one night of restaurant food, clean sheets, and hot showers. There I had a surreal experience. Hungry for news from the US, I walked around to a news stand. The *Newsweek* I bought had block headlines reading something like "DETROIT BURNS." That was the summer of 1967, "the long hot summer" when some 160 race riots erupted across the US. To me, standing on a sidewalk in Tegucigalpa, it seemed like my country was self-destructing.

Our last major site was over mountains from Tegucigalpa to Danlí. We set up our camp a mile or two out of town, which in hindsight was too close. Downslope several miles was a Don Thomas cigar tobacco farm. The company had moved there after the Cuban Revolution. The farm manager had been educated in the US and brought me good cigars for the opportunity to talk in English. He told us that a widespread rumor in nearby Danlí was that we were prospecting for gold—

obviously, their gold. I didn't know if gold had ever been found in that district, but to the natives we sure looked like prospectors. Eduardo tried to explain whenever he could, but our story of being bat collectors apparently was implausible to the poor citizens of Danlí. Nothing came of their suspicions and potential hostility, but the situation was discomforting.

One of our last efforts was at Chichicaste, downslope from Danlí and close to the Nicaraguan border. There, among many finds, we caught two more wrinkle-faced bats and a velvety fruit-eating bat.

The trip was nearing its intended end in late August, but there was one more pot bubbling. Richard had not felt well much of the trip and spent an uncharacteristically long time lying in his cot. At Danlí, the situation was finally too serious to further ignore. We left Eduardo at the camp and I drove Richard over the mountains to the American embassy in Tegucigalpa. A doctor gave him some antibiotics and recommended he immediately fly back to the states. He did, and I drove back to Danlí.

Eduardo and I hastily packed camp, including two dozen wet, fresh bat skulls. We drove again to Tegucigalpa where I paid Eduardo his wages and bought him a bus ticket to Comayagua. At the embassy, I called Dr. Carter at A&M. Here I was, 22 years old, alone with no Spanish comprehension, and with four countries (Honduras, El Salvador, Guatemala, and Mexico) between me and Texas—oh, and with firearm weapons permitted only in Richard's name. I was told to bring those weapons back, so being easily intimidated at that stage in my life, I did just that. Later, I decided never again to repeat such a dangerous, bone-headed move.

I telephoned A&M every night on the way; the wildlife faculty were keeping track of my progress, presumably with their fingers crossed. My technique for getting through international borders with illegal guns was to hide the guns under field equipment boxes just behind the seats and to roll up the windows so the rotting wet bat skulls would stink up the truck. It worked. The border guards were disgusted and soon despaired of any more thorough inspection.

Remember that Civil War in Guatemala? It was still going on. I don't recommend running guns through a Central American revolution.

When after three days I passed into Mexico, I felt like I was home free. But all my problems were not over. The first was easily solved. How can you possibly find your way driving through Mexico City? My solutions were to wave down a taxi and offer the driver $20 to lead me through the city to the highway for Queretero, on the route to San Luis Potosí. He agreed. Twenty dollars in Mexico in 1967 was some real money. We barreled through those impossibly marked race tracks called streets, me right on his back bumper, until on the north side he gave me a high sign and waved me on.

I made it to San Luis Potosí before my money was gone. What I did have was an emergency check for $100—a cashier's check on a College Station bank. I was turned down at more than one bank before a kindly bank officer who spoke English came to the lobby and explained the problem. There was no legal banking agreement between Mexico and the US, and therefore a cashier's check from a US bank was unguaranteed and worthless to him. He recommended I look for help at the American Consulate. The Consular General saved me. He cosigned the check and sent me to his banker who accepted the check, so I was able to head for Texas.

I pulled up to the A&M University wildlife department 5 days after leaving Danlí. My first expedition into the wilds of Latin America was complete, I was okay, the specimens were all safe and available to add knowledge to the bat fauna of the Central American core, and, best of all, Richard was recovering.

Uniformed Interlude

Eight months after returning from Honduras, my educational plans were put on hold by the Tet Offensive in Vietnam. Like many other students, my graduate school draft deferment was canceled.

In the spring of 1968 I passed the draft physical exam and so elected to enlist in the US Air Force for a four-year stint. I graduated at A&M in May 1968 and headed to basic training in San Antonio in July. Under most circumstances at that time, someone with a college degree would enter officer training. But with the draft operating in Vietnam mode, the military services had far too many officer candidates, and they certainly did not want officers or even enlisted men in the fields of ornithology or wildlife science.

They sent me to Defense Language Institute, West Coast Branch, at the Presidio of Monterrey, California where I spent most of eight hours a day for six months studying Spanish—unfortunately too late for the bat expedition to Central America.

The Presidio is on the Monterrey Peninsula that projects into the Pacific Ocean—John Steinbeck territory. Offshore, below the Presidio, lived sea otters, sea lions, and the occasional puffin and gray whale. For a man from the dry brushlands, being there was a splendid nature excursion. I even learned to drink wine with a meal. Coming from Bible belt Texas in those days, I assumed the only legal adult beverages stronger than water for any post-breakfast meal were coffee or iced tea. Of course, I didn't get a lot of opportunity to try my new taste for wine; an airman's monthly salary in those days was about $50. Fortunately, with west coast wildlife within walking distance, my entertainment expenses were light.

After completing language school, one of my life's great ironies occurred. The Air Force sent me for further training to Goodfellow Air Force Base in my hometown of San Angelo. I was stationed there for about eighteen months and then was off to Homestead AFB, south of Miami. For most of my time at Goodfellow, despite being single, I was allowed to live off base. It was a highly unusual privilege that allowed me to continue studying Concho Valley bird life. I also then began my life-long association with Angelo State University by starting their scientific bird collection. When not busy at the base, I collected and prepared scientific specimens of birds.

What I was trained to do for the Air Force was interesting and important, and I learned a language that served me the rest of my life. In my Air Force profession, I flew as a voice-processing specialist in the back of a C-130 Hercules from Homestead and later an RC-135 from Offutt AFB in Omaha, Nebraska. We flew up and down the north coast of Cuba. My job was to listen to radio communications of pilots in the Cuban military. Other than for the air-borne refueling of our RC-135 over the Florida Peninsula, I never thought of it as dangerous. I knew it was important. Only eight years prior to my small contribution, there occurred, early in the presidency of John F. Kennedy, the scariest event in the Cold War: the Cuban Missile Crisis.

Southern Florida was another naturalist's dream with American crocodiles and alligators, West Indian manatees, tiny Key deer, snail kites, smooth-billed anis, roseate spoonbills, and clouds of egrets and herons. I regularly drove down the keys and joined party fishing boats that went into the Gulf Stream. Those trips led to one of my earliest scientific publications, a note on gannet migration off the east coast of Florida. The highlight was when I managed to secure a pass to join a National Park Service team banding sooty terns at Fort Jefferson in the Dry Tortugas islands, sixty-eight miles west of Key West in the Gulf of Mexico.

When the time arrived to complete my military service and return to the civilian life of a naturalist, the Air Force wanted me to consider a career with them. The truth is that they spent a lot of money training me, and they were good to me. An example of just how good occurred when I narrowly missed an Article 15 disciplinary action, just short of a court-marshal. At Homestead AFB, I missed a mission flight because I was out in Everglades National Park bird watching.

I forgot the flight. They were not impressed with me. I wasn't impressed with myself, and I'm sure I deserved punishment beyond the reprimand I received. But it does illustrate why I went my own way. Becoming a professional naturalist had always been my dream

Eagle Trouble

Upon completion of my Angelo State University master of science degree in 1974, I headed back to Texas A&M for an attempt at a doctoral degree in Wildlife Science. Keith Arnold had agreed to take me, he acting as chair of my graduate committee, and I felt good about that, but I did not consider this degree attempt to be a slam-dunk as coursework in statistics was required. Keith and I agreed on a research topic— historical change in the bird fauna of the San Angelo region of the western Edwards Plateau. Included in the investigation would be a thorough quantitative description of current bird communities in the various ecological settings. And so following successful completion of coursework, including the dreaded statistics, in fall 1975 I headed back home to begin field work.

Describing bird communities requires a seasonal population census of each species in each community. I used a technique, variable width strip transect, that required walking a route (avoiding rattlesnakes) through vegetation while recording each individual bird detected and estimating its distance from my transect line. For every study plot, each representing different vegetation, I had to repeat the census several times within both breeding and winter seasons. My plots were located within an area of about fifty miles west to thirty miles east of San Angelo, so there were many early mornings with long drives to reach sites before sunrise. It was a lot of work.

During my first winter of field work, I lost use of some hard-earned data because I was evicted from a ranch long before the study was completed. That eviction is the subject of this chapter.

In the world of bird terminology, a raptor is a bird of prey and therefore typically some variation of hawk or falcon. The largest hawks are called eagles. Among the world's eagles are a group of about sixteen species in the genus *Aquila*—the so-called booted or typical eagles.

Included are the wedge-tailed eagle of Australia, steppe eagle of southern Europe to Mongolia, and golden eagle through much of the northern hemisphere including Texas. With maximum wingspans around seven feet, they are massive and most often are the largest raptors wherever they occur.

Booted eagles are capable of killing wild prey as large as coyotes and adult deer or even pronghorns, but numerous diet studies show that their normal prey are such mammals as cottontails, hares, and ground squirrels. On the other side of the Atlantic, birds like grouse and pheasants are commonly taken. The Berkutchy eagle falconers of Eurasia use them to hunt foxes and hares. The problem arises where booted eagles occur with pastoral people. There the birds will prey on domestic livestock. The eagle/livestock conflict has been intense in Scotland, Australia, and western North America, all areas where sheep production on rangeland has been a major industry.

Golden eagles occur throughout North America but are at their greatest density in the mountainous west in Canada and the US where they favor nesting on high cliff faces. The species is migratory, moving southward as winter conditions develop. Most that occur in Texas arrive in October to December and are largely gone back north by the end of March. One recent evaluation estimated a winter population south of Canada of 63,000 individuals. How many of those spend winter in Texas is hard to even guess, but it surely is only some fraction of that. There are a few pairs that nest in spring and summer in Texas from the lower Pecos area near Del Rio west into the Trans-Pecos and on the Caprock Escarpment bordering the High Plains.

From the time of my childhood I heard them referred to as Mexican eagles. I am not certain about the origin of that regional name but presume it was from an assumption that eagles arriving in Central Texas in the fall are coming from Mexico and the Southwest or that the bird's major occurrence is in Mexico. The reality is that a Golden Eagle in Texas in January is far more likely to have come from nesting in Alberta or Wyoming than from Coahuila.

My hometown of San Angelo was known during the first half of the twentieth century as the inland wool capital of the world. The

Figure 3 golden eagle

shallow soils and rocky slopes of Edwards Plateau rangeland are optimal for domestic sheep and goats. The climate and ecology of the region as well as market forces generally dictate that birth of lambs (lambing in local parlance) occurs in winter. And so there you go. Newborn lambs are at the age that makes them most vulnerable to predation. Unfortunately, lambing then coincides with arrival of migrant golden eagles setting up what was one of the most intense of western wildlife conservation conflicts in US history. San Angelo was the epicenter of this conflict in Texas, and I was so lucky to be a resident ornithologist.

When I was young and filled with passion for bird life, I was predisposed to accept the often-stated position of bird conservationists that golden eagles did not prey on livestock but rather wholly on native animals. What I know today is that golden eagles at times in winter aggregate in unusual numbers on at least a few ranches where they indeed prey on enough lambs and kid goats to be an economic problem for the impacted ranchers. But what is also clear is that in most places with livestock production such an extreme situation does not exist; eagles more typically are locally few, widely dispersed, and not a problem. It also is undeniable that lambs die from many causes: failure to thrive, stillbirth, injury from extreme weather events, starvation and predation from many carnivores other than eagles. Also, eagles will eat lamb carrion when they are not responsible for the lamb's death. It is not my intent in these pages to defend golden eagles. I am not an expert on eagle/livestock interaction, and each side of this dispute has argued its case many times. Rather, the point here is what happened to me when in my naivety I stepped into this issue without any intent to do so.

Predators often are reviled for simply being what they are, regardless of what damage they might or might not do to private property. As a youngster looking out the window of my family's car, I often saw dead hawks and owls hung up on roadside fences as a display of justice meted out to predators. Never mind what was the actual diet of the offending raptor, such as Swainson's hawks that eat predominantly grasshoppers in fall migration through Texas and ground squirrels, other rodents, rabbits and snakes in breeding season.

In that spirit, in the third quarter of the 20th century, eagles were hunted from airplanes. It was sport for some but predator control for others, even though golden eagles had been for most of that time federally protected by amendment of the Bald Eagle Protection Act of 1962. During my first year on the ranch in question, a famous court case began in San Antonio. Between December 1975 and January 1977, approximately seventy-five golden eagles were allegedly shot from a helicopter in Real County (which incidentally, if true, is an example of an unusual aggregation of eagles that could have caused some real damage). The offenders came to trial, much to the anger of the livestock industry. The case made national news and was reported by legendary TV newsman Walter Cronkite on CBS. The accused were convicted but received little more than slaps on their wrists.

Okay, so here I was on this sheep ranch surveying mostly songbirds. I never saw an eagle or an airplane over the place. Another student from another university was doing a wildlife study on an adjacent ranch. That other student overheard a conversation indicating the hunting of eagles from aircraft. The rest of the story is that upon completion of his study, he reported the alleged conversation to federal authorities who began an investigation. To my knowledge, nothing came of the investigation other than that I got kicked off the ranch on the presumption that I was involved in the report to the feds.

I'm tempted to argue that I did nothing wrong and I suspect the rancher thinks I did very little right. Probably neither position is fully correct. Someone may have broken a law by killing eagles, but then I clearly broke laws of 4 countries by transporting improperly permitted weapons through Central America in 1967. It's not like I'm sainted and can piously cast stones. There must be more than one moral to this tale, and I am not certain that I'm on the right side of any of them.

The story of my supposed transgression worked its way around the ranching culture, and I could easily have been evicted from bird research on any and all private property in this inland wool capital of the world. I could have lost my entire doctoral research program. Had that occurred, I would have had to leave for new territory to continue the practice of my trade. It did not happen. Oh, I am denied access to some property, and in at least a couple of cases I know my eagle trouble is the reason, but most denials to my requests for entry are mistrust of biologists in this era of the endangered species act of 1972. Being a working naturalist in West Texas has its pitfalls.

The eagle issue is quietly on the back burner in Texas these days. I almost never hear anything about them, and if any are being killed, those with knowledge of it have their mouths firmly closed.

Quebrada Vainilla

In 1983 it had been the better part of ten years since my last tropical experience, and I was feeling the loss, so I called John O'Neill. "Do you or anyone else at Louisiana State University need any bird collecting help this coming summer?" They did, in Peru, and I went. Whether or not they really needed my help I don't actually know, but they at least let me help. I was and remain deeply in their debt. It was to be my first trip to South America, the bird continent.

There are not enough superlatives to describe the natural history of Peru. Roughly one-fifth of the world's bird species have been recorded in that Andean-Amazonian nation. Mammals include almost thirty species of primates along with spectacled bear, jaguar, and probably more than 130 species of bats, and it is a botanical wonder with almost 14,000 known plant species.

O'Neill was a student at the University of Oklahoma when he travelled to Peru in 1961. On that trip he made a small collection of bird specimens, but George Sutton, the legendary Oklahoma ornithologist, told him he should work with George Lowry at LSU if he wanted to do collection-based work in South America. So that's where he went. He flourished at LSU, found funds for field expeditions, and concentrated his efforts on the bird fauna of Peru. With the addition of Van Remsen to the LSU team in the late 1970s, exploration of the Peruvian bird fauna occupied well over half a century of effort.

Our 1983 field work was in Amazonia, east of the Andes Mountains. We worked on the east bank of Quebrada Vainilla (Vanilla Creek), a tributary of the Rio Amazonas. Its mouth is only about seven miles upstream from the mouth of a large Amazonian tributary, the Rio Napo.

Our camp was in *terra firme* rain forest. At the broadest ecological scale, Amazonian lowland forests are separable into *terra firme* and *varzea*. *Varzea* is seasonally flooded forest close to stream channels. *Terra firme* is high enough, by at least a few feet, to not be flooded in the wet season.

I flew to the Peruvian capital, Lima, in early July then hopped a domestic flight across the Andes to Iquitos, Peru's largest Amazonian city. There, I met the team who were resting and resupplying between field sites. As usual in the Amazon basin, travel to the quebrada was to be by boat, and just getting such a vessel outfitted and loaded is no minor task.

Shopping for groceries means visiting an open air mercado. You can find modern supermarkets in a city the size of Iquitos, but most of what you need is available in the traditional mercado and it's cheaper. Cheap is important for managing the scant funding of a field expedition. The shopping list includes dried grains (rice, beans, oatmeal), powdered potatoes, small bread loaves, peanut butter, canned milk, canned butter-like product, canned tuna, and actually just about anything else canned. If you can find Tabasco sauce, grab lots of bottles of that. Out in the field, you crave any intense taste. While you're there, you can get some fresh vegetables (cabbage, carrots, potatoes) for the first few days afield before it rots in the tropical air.

The Amazon flow fluctuates enormously between seasons, even as far upstream as Iquitos. In the wet season, the river is already two miles wide there, and Iquitos is still 2300 river miles from the Atlantic Ocean. July is the dry season, so supplies to be loaded on our boat had to be shouldered down a long wooden stair to river level. I was reminded again how my back is different from a man who labors with his. Abraham Urbay hefted an outboard motor down that stair. I would guess the steep stair at that point in the dry season was no less than seventy-five steps and the motor at least 150 pounds. I have no idea what the expedition was paying the man, but it wasn't enough.

Finally loaded, we motored downstream to the mouth of the quebrada and then up the creek about three miles. Our *terra firme* forest camp site was not far from the Vainilla, and more good forest habitat extended inland to the east. There were good hunting trails to work.

The expedition leader was Steven Cardiff, an LSU graduate student. The full team included additional graduate students Donna Ditmann and Angelo Capparella, undergraduate Tristan Davis, two native Peruvians: Manuel Sanchez and Abraham Urbay, and me. From several aspects, an essential member of our group was Manuel.

Manuel is ethnically Quechuan, an Andean native group descended from the Pre-Colombian Inca Empire. About 1968 he was "discovered," actually recommended, on the basis of his skills as a carpenter and all around handy man for employment with a bird expedition led by John Terborg. Reportedly, Manuel couldn't imagine any practical or even sensible use for scientific collecting of birds, but he needed the employment. For about four subsequent decades he was a highly sought-after member of most Peruvian and many Ecuadorian bird expeditions. He was in charge of building field camp in whatever setting, rainforest to alpine tundra. He set up and operated the mist net regimen and became skilled at preparation of scientific specimens. By the early 1980s, his wife, Marta, was with most expeditions as well. She cooked, fished and tended camp in general.

When I showed up in Iquitos with my small tube tent, I made sure I had a rain fly to stretch above it. I knew that it rained a lot, even in the dry season. That turned out to be a laugh. I had no idea the preparation necessary to stay dry in Amazonia for a month. As soon as we arrived at our camp site, Manuel proceeded to cut sapling trees for stout palos (poles) with sharpened points. He pounded the palos into soil; four were corners of a square about thirty feet on a side, two were tall end poles, and one was an extra-long ridge pole. To each corner palo he attached the corner of a huge sheet of industrial thickness plastic. Each corner palo was then further secured with rope to a nearby tree. I witnessed construction of a storm-proof plastic-roofed campsite. He then built a cooking table, complete with lower shelf to store food stuffs off the ground. The rest of us machete-chopped the undergrowth from our small city, erected our personal tents and the large work tent under the plastic, and dug shallow drainage trenches around the plastic perimeter. Manuel finished up by building a plastic antechamber to the work tent, an arrangement that forced insects, like mosquitoes, to get through two doors and an insecticide-fogged ante-chamber before finding human flesh. I was to become humbly grateful for all this construction. Mosquitoes (hosts for malaria and yellow fever) and sand flies (that transmit *Leishmania*) were a nocturnal fear, and it rained much more than a lot. The protozoan parasite *Leishmania donovani* causes the terrible disease leishmaniasis that was treated at the time with antimony-containing drugs. Antimony is a heavy metal, the injection of which is not pleasant.

In 2011, Manuel and Marta were honored in a manner much revered by organismal biologists—a bird species was named for them, *Turdus sanchezorum,* the varzea thrush. The type specimen for *Turdus sanchezorum* was collected at Quebrada Vainilla on this very same 1983 expedition in which I was a participant.

Working in the rainforest can be dangerous. It's usually cloudy and with the density of trees, sun direction is obscured on even a sunny day. One can easily lose direction awareness. If I shot a specimen and it fell to the forest floor, I had to leave the trail (which wasn't much of one anyway) for perhaps dozens of yards, and search in widening arcs for the bird lying among leaves. Finding the bird, you then look up and realize every direction appears the same—trees, trees, and more trees. The trail is thirty yards in one direction and then 2000 miles to the Atlantic in another. Getting lost is avoided by marking your search route with plastic flagging.

Another concern is venomous snakes, various arboreal vipers and then fer-de-lance and bushmaster in particular. Curiously, I never encountered any of them in Peru but did later in Ecuador and Guyana. Looking up for birds precludes looking down for vipers. Naturalists have been bitten, but fortunately, never on trips on which I was a participant.

I had experiences with mammals in that forest that I never repeated elsewhere in the American tropics. One day I followed a giant anteater for a few yards, leaving it when in apparent awareness of being trailed, it turned back toward me. On another day, a small pack of bush dogs crossed my trail. This short-legged forest dog is little known and rarely encountered.

Quebrada Vainilla was my virginal dedicated experience with

neotropical forest birds. My first, and lasting, impressions were that: they are much more difficult to see than birds in Texas brushland, parrots and especially tinamous can be the most difficult of all to see, there is no telling how many different tanagers exist in any locale, and suboscines are the most compelling of Amazonian birds.

A popular notion is that all or most tropical rainforest birds are brightly, even garishly, colored, and some are. Many of the parrots, hummingbirds, barbets, cotingas, and tanagers, for example, are riots of color. But the floor of the rainforest can be dark to the point of foreboding. I grew up seeing the blue horizon, and I was surprised to find how much I missed that after a month in tropical forest. Color is of little use in that deep shade. So there you find many of the somberly feathered antbirds, one clan of the suboscines.

Figure 4. Shaded floor of the tropical rainforest

I should explain suboscine. In avian classification, the largest order is Passeriformes, the perching birds that make up over half of the world's roughly 10,000 bird species. The order is divisible into suboscines and oscines. The oscines have more elaborate anatomical vocal mechanisms and often deliver more complex songs; we often call them the "songbirds." Our familiar mockingbird and cardinal are oscines. Suboscines are fewer and most of them are in South and Central America. My home state of Texas hosts, at least fairly regularly, about thirty species of suboscines (all of them flycatchers) and another thirteen have been seen a handful of times. On the other hand, the region from Mexico through South America has recorded over 1100 suboscines. Fortunately, we had our share at Quebrada Vainilla.

Finding short-billed leafscraper, black bushbird, and chestnut-shouldered antwren were good examples of one goal of such expeditions. Our site was for each of these three suboscines only the second locale known for them in Peru. In other words, we put some important dots on bird distribution maps.

In Texas, woodpeckers are the common tree trunk climbers, and they fill that behavioral role in the tropics as well, but there's also a suboscine group of tropical tree climbers: woodcreepers. They're invariably brown and gray bark color with pearl or buff-colored streaks, dots, or teardrops. I was intrigued with them and often found my eyes wandering from one tree to the next in search of woodcreepers.

Ground-dwelling birds can be particularly difficult to see. Specimens of variegated and white-throated tinamous were taken by our team, but to this day I have never seen a living tinamou in the wild. I heard them often, just never saw one.

Texans are familiar with North America's impressive ground-cuckoo, the greater roadrunner. It's famed for its running speed, diet of lizards, snakes and whatever else it can overpower, and for its stardom in cartoons. Probably, most people within its range don't realize that the roadrunner is a legitimate cuckoo, and there are several more large ground-cuckoos in the American tropics. One almost ran me over in the forest. I heard an unfamiliar sound in dense, ground-level foliage, so I squatted on my haunches and peered in the sound's direction. Then in a split second a ground-cuckoo, running full out, came directly at me and bounded past, not two feet from me, giving me no chance at a shot. Manuel did secure a specimen of red-billed ground-cuckoo.

There were, of course, plenty of the more colorful tropical birds. curl-crested and ivory-billed aracaris, part of the toucan assemblage, were regular along the quebrada. Two birds that fairly glowed in their shiny green plumage were the pavonine quetzal and great jacamar.

A typical day in such a field camp involved collecting and specimen preparation. Collecting, with weapons and mist nets, was a morning activity. Everyone was up before sunrise. If I got up early enough there was time for a bowl of oatmeal, but if too slow then I just marched into the forest with my shotgun. Nets were also opened before sunrise.

Without refrigeration to prevent specimens from decomposing in the tropical climate, a firm rule was to not collect more in a morning than could be prepped by the end of the evening. Typically, everyone was back in camp by noon mealtime for a culinary delight of noodles and canned tuna fish. The unquestioned essential condiment in camp was the biggest available jar of industrial strength Tabasco Sauce. We craved flavor. In that climate, fresh vegetables and meat would hardly last two or three days so our store of food for a month was dried or canned. The following year when Marta Sanchez was with us, she would on some days catch fish in the morning, and we feasted.

At the end of our noon meal, those who would prepare specimens retired to the work tent where much of our field lives were lived. In the tent was a folding table that could seat four to six people. To be useful and therefore take up scarce work space and resources on such an expedition you need to be able to prepare professionally two or three birds per hour over several hours duration. A gas lantern sat in the middle of the table. Manuel built a two- to three-foot tall wooden tower positioned above the lantern. We laid freshly stuffed bird specimens on tower platforms such that the rising heat from the lantern dried them.

A treat we allowed ourselves in the work tent was music. This was the era of battery-powered portable cassette players. Each team member was allowed to bring his or her favorite music. My tastes are unsophisticated and eclectic, so I was prepared to enjoy or put up with all of it until Angelo played Pink Floyd. Listening to *The Wall* when you're dog tired in the dark of the rainforest was depressing or even agonizing. I gave him some grief, but it was mostly light-hearted.

It was not uncommon to spend ten or more hours a day in that tent, and clearly to all involved, the enthusiasm for that grind derived from a personal commitment to the value of specimens in the effort to fully understand a bird fauna. I am aware that many who read these pages with interest in natural history or birds in particular object to scientific collecting—killing birds to gather the sorts of information achievable in no other way. I am inadequate to make any new, more useful defense of this important contribution to the zoological inventory of Earth, but others have done so and done it well. If you wish to read justifications for scientific bird collecting, please consider Remsen, J. V., Jr. 1995. *Bird Conservation International*, vol. 5, pp. 145-180; O'Neil, J. P., D. F. Lane, and L. N. Naka. 2011. *The Condor*, vol. 113, no. 4, p. 879; and http://www.universityofalaskamuseumbirds.org/the-growing-power-of-the-ranks-of-the-dead/.

On another much less serious note, many who will read this treatise are not likely searching for scientific information; this treatise is, after all, about my experiences more than about the birds, bats, and snakes. So I use here mostly English names rather than scientific names for species. In my professional life, Latinized scientific names are standard and more informative, and many scientists tend to more or less dismiss English names as barely necessary. But they are important to many of my readers and remain so to me in some contexts. I state all that because this is my first opportunity to express a personal opinion in print about bird English names. If you don't want to consider my peeves about bland names, then move past the next paragraph.

Names should be informative but there is no reason that I can imagine for them to be uninteresting. In the US we have eastern this, western that, northern and American whatever, and so on. I admit there is geographic logic to these compass and political appellations, particularly where there are species pairs or larger groupings. But, with few exceptions, North American bird names are bland, sanitized of interesting descriptors, poetry, history, or even humor. My least favorite names are those that begin with the word common. One of the least common hawks in the US is the common black-hawk. I bring this subject up because many neotropical birds have been given wonderful names. Some of my favorites are hyacinth visorbearer, bearded mountaineer, and marvelous spatuletail among the hummingbirds; and Roraiman antwren, Jocotoco antpitta, trilling tapaculo, creamy-rumped miner, and Vilcabamba thistletail among the suboscines. A sincere "thank you" to those that dreamed up these entertaining and memorable names.

We left Quebrada Vainilla with 1164 specimens, having prepared an average of about forty-nine a day. Everyone was tired but in good health and longing, I am sure, for a hot shower and a cheeseburger. The major task remaining fell to Abraham Urbay, who now had to fight gravity in shouldering that 150 pound boat motor up what now were probably 100 stair steps.

I returned to Peru in mid-July of the following year, 1984, to again assist an LSU expedition, this time to an island in the Amazon. Isla Pasto is about fifty river miles downstream from Iquitos. We were given logistical support, boat transportation from Iquitos to the island, by Explorama Lodge. Angelo Capparella led the team, which included me, herpetology graduate student Brian Smith, and three Peruvians: Manuel and Marta Sanchez, and Robinson Chota.

Camp and daily activities were run pretty much as the year before, except for the welcomed contributions of Marta Sanchez. Manuel was anxious to not leave her alone in their Andean home. The Sendero Luminoso (Shining Path) was active up there. Senderos are Maoist guerillas noted for their brutality, and although they remain marginally active in Andean regions, their terrorism is greatly reduced from what it was in the early 1980s. Marta gladly came with us.

I brought a box of cigars on this trip with the intent to sit well out of camp after lunch every day and enjoy a cigar. Early on I noticed Marta watching my routine with interest, so I offered her a puro (cigar) which she accepted with enthusiasm. Although my cigar supply dwindled more rapidly, the companionship was enjoyable.

The summer on Isla Pasto was for the most part routine for one of these expeditions. The mornings were for collecting specimens and we occupied the afternoons and evenings with specimen preparation. Of note on this island were Amazonian river bank specialists such as short-tailed parrot, tui parakeet, and plain-breasted piculet.

I never in all my South American summers lost fascination with piculets. They're the smallest of New World woodpeckers. Back in Texas, several of my graduate students studied woodpeckers for their thesis research. One of my students, Michael Husak, and I authored the species account of golden-fronted woodpecker in *The Birds of North America*. Being immersed in woodpecker biology, I was then constantly fascinated with encountering these tiniest of the group.

One day, Robinson Chota brought into camp a horned screamer.

Figure 5. horned screamer

For a teaching ornithologist, screamers are a common classroom story. They have the posture and size appearance of turkeys. Even the bill looks superficially like a galliform (chicken-like bird). But they decidedly are not of that clan. DNA and certain anatomical features place them firmly in the waterfowl order with swans, geese, and ducks. Waterfowl evolutionary origins extend well back into the Cretaceous, over sixty-five million years ago, and the screamers, looking nothing like a duck, probably are an early branch in that tree. This particular species has an odd feather shaft arising from the crown of the head—the "horn."

It was on a day excursion to Isla Correviento, about three miles downstream from our camp, that we encountered the world's largest suboscine, a species of continga called the Amazonian umbrellabird.

Figure 6. Amazonian umbrellabird

They're black. The male is almost two feet long and not only sports an umbrella-like crest but has a long feathered wattle hanging from his breast. The world has a nice selection of oddly adorned birds, but umbrellabirds are right up there with the best of them.

Our return to civilization was again facilitated by an Explorama boat, and this time we stayed overnight at their main tourist lodge. It was odd to be wined, dined, and pampered in the Amazonian forest, a place more familiar to us as a primitive work zone. No one complained.

Then, as was my place in life, I flew back to the brushlands of the Texas southern plains. At this time in my career, I was the head administrator of the ASU Biology Department. I taught class, advised colleagues and students, attended long meetings, filled out interminable forms, and attended to my corner of the academic bureaucracy, all the while dreaming of Amazonia and piculets.

North Slope

In the late winter of 1985 I was feeling sorry for myself again, for having no exotic prospects for practicing my field trade in the coming summer. In that frame of mind, I was reading an American Ornithologists' Union newsletter where a notice caught my eye. The US Fish and Wildlife Service in Alaska requested volunteers for the Tundra Bird Project on the North Slope of the Arctic National Wildlife Refuge. Well, why not? As a seasoned hand now with the tropics, I might as well experience a radically different climate and fauna.

Much to their surprise, I suspect, I applied and was accepted. They were not anticipating someone with a Ph.D. to apply for an unpaid volunteer position. It was decided that I would lead one of seven field camps, and when the spring semester at ASU ended in mid-May, I headed for Fairbanks.

The Arctic National Wildlife Refuge (ANWR) has been much in the news for more than three decades. The most northern portion of ANWR is on the coastal plain bordering the Beaufort Sea, the famous North Slope, reputedly with huge petroleum deposits. To the west of the refuge is the oil famous Prudhoe Bay, head of the Trans-Alaska Pipeline. Also on ANWR are the calving grounds of the largest caribou herd in North America: the so-called porcupine herd that winters south of the Brooks Range. One can see muskoxen, wolves, grizzly bears, snowy owls, and gyrfalcons. For most Americans, the fame of ANWR, of course, comes from the controversy of whether to drill for oil.

As it is delimited today, ANWR was created by the Alaska National Interests Land Conservation Act of 1980. Among the Act's provisions was a requirement to conduct a comprehensive coastal plain inventory of fish and wildlife resources. Part of that inventory effort was the Tundra Bird Project.

In Fairbanks we provisioned ourselves for the climate north of the Arctic Circle. To a man who grew up close to the Chihuahuan Desert, this was a fairly critical activity. Then we spent a full day in a sobering bear safety program. I had never been on foot around wild bears before, and since my mother had not raised a fool, I paid attention to the instruction. Only grizzly and polar bears (no black bears) occur on the North Slope. The grizzlies up there are small; the polar bears are not. One of the refuge hard rules was "don't approach bears." You would think that went without saying, but I was to learn otherwise.

The only excitement we enjoyed in Fairbanks was with a dog-tired smokejumper at Eielsen Air Force Base. All the bird project volunteers were housed in barrack rooms at Eielsen for our days of orientation. As the 40 year old elder of the group, I was semi-in charge of room assignments—at least that's my story. Two of the volunteers awakened me late one night with a crisis on hand. There was one really angry firefighter standing in the door of their room. Someone had mixed up room numbers. This guy had been jumping out of airplanes to fight fires for two exhausting weeks, and he was not amused that a bird watcher was in his bed. It did not require higher order decision making to let him have the room.

We flew out of Fairbanks in a goony bird, the legendary DC3. Dating back to 1935, this venerable airplane was a stalwart of commercial aviation and transport in WW II. It was a workhorse in flying supplies over "the hump" from India to China to supply the war effort of Chiang Kai-shek. From the looks of it, our plane was that old. Our destination was Peter's Lake in the near treeless Brooks Range that stretches east to west across northern Alaska. Even though it was early June, the lake was still frozen deep enough to support an airplane with about twenty-eight people on board. We landed. What I learned later was that the pilot had not before landed on this patch of lake ice, boxed in by mountains. Our group was an ignorant participant in his landing experiment.

Peter's Lake was a staging area for helicopter transport of each of the seven teams to their respective research sites on the tundra. The sites were given wonderful native-inspired names like Katakturuk, Sadlerochit, Okpilak Delta, Niguanak, and Jago Bitty. My team's camp site, Aichilik, beside the glacier-fed river of that name, was the most easterly, the closest to Canada. Camp consisted of a work/cooking tent, a supplies storage tent, and our individual sleeping tents.

Figure 7. frozen Peter's Lake, Arctic National Wildlife Refuge in Alaska

Figure 8. camp on the North Slope near the Aichilik River.

Birds at this site had been studied the year before, 1984, and our effort was follow-up. I had a team of two young men, Douglas Braddock and David Whiting, who were recent graduates. Fortunately, David had worked his way through college as a fry cook.

Our task was to survey breeding bird populations on different tundra habitats. Although incredibly simpler than environments in my native Central Texas, the tundra can be, for example, wetter versus drier and with or without willow shrubs. Birds do respond to those seemingly minor ecological variations.

Our daily schedule was to get out in the morning onto one or more of our fifteen surveyed plots where we walked three abreast along transects, identifying flushed birds and searching for nests. Walking was done in rubber boots as the ground is mostly sloppy wet. A stiff breeze consistently blew in from the Beaufort Sea. The breeze was most welcomed for an infamous Alaskan reason: mosquitoes. When censusing, one longed desperately for those transect walks facing the wind. Mosquitoes by the dozen would then swarm the back of your head, biting only your ears. When you turned to walk with the wind, the mosquitoes also turned to your face where they especially enjoyed your nostril and ear holes, wet eye margins, and actually, all the rest of your facial skin. Mosquito netting was not a good option because it obscured the observation of birds.

If a grizzly was on the plot we were to walk away and let the bear have it for as long as he wished. For protection from bears, the refuge officials issued us a twelve gauge shotgun with slug and firecracker shells. The officials preferred that we not kill their bears; we were, after all, on a wildlife sanctuary. But if a bear charged, we were instructed to fire slugs into its shoulders in an attempt to break down those skeletal joints which hypothetically would stop it from running at us. Uh, yeah.

Toward the end of summer, one of my team got bored and broke the most important bear rule. He approached a grizzly hoping for a chance at a better photograph. I was in the work tent when I heard shouting outside. Looking out of the tent, I saw the man running full out with a grizzly loping behind. He dove into the tent. I grabbed the shotgun and loaded some firecracker shells. The bear's reaction to shells exploding in front of him was not the one we wanted; he stood his ground, shaking his head with slobber slinging everywhere. Oh brother.

We crawled out the back of the tent, crossed the shallow Aichilik, and waited on the opposite bank ridge. The grizzly got into both large tents and tore up some supplies and all my scientific specimens of lemmings but thankfully left the two-way radio unharmed. After what seemed like hours, he wandered away and we were able to get to the radio and call for help. Refuge headquarters sent the helicopter, but the bear was long gone. We mounted a guard for about forty-eight hours but had no more trouble.

Wildlife is highly visible on the tundra. The caribou, visible from miles away, rather calmly walk right through camp in their unceasing movements. Unfortunately, we only once had a muskox within view.

The bird fauna, our research objective, was an odd mix for my experiences. One glaring example is that there are very few songbird species in the Arctic. Lapland longspurs, savannah sparrows, and hoary (Arctic) redpolls were most of what we saw of that largest world order of birds. But for a shorebird enthusiast, the North Slope is Nirvana. The conspicuous nesting species on our plots were: American golden-plover, ruddy turnstone, semipalmated sandpiper, pectoral sandpiper, stilt sandpiper, buff-breasted sandpiper, and red-necked phalarope.

Figure 9. nesting birds, North Slope tundra

We found one nest of whimbrel, the first breeding record for the refuge coastal plain. Pomarine jaeger , long-tailed jaeger, golden eagle, and short-eared owl were the regular avian predators. I was thrilled to spot two gyrfalcons.

Although you might think otherwise, especially without many songbirds, the tundra in summer is not often quiet. The most conspicuous sound on our stretch of it was the incessant clucking of willow and rock ptarmigans.

Figure 10. Willow ptarmigan

One morning, a ptarmigan startled us by landing squarely between one of my group's feet. Moments later a gyrfalcon zoomed over our heads. Clearly, for this ptarmigan, we were the lesser threat.

Refuge headquarters was at the native Inupiat village of Kaktovik on Barter Island. Back in Texas, I was naively excited about the prospect of meeting the Inupiat people, but the result was disappointing—it didn't really happen. They seemed not to want to meet me. It appeared that they considered those of us from the lower forty-eight states as interlopers who took refuge employment away from them. When the refuge was established, the Inupiat lost a lot of their hunting grounds—an old American story.

At the edge of Kaktovik, resting on the beach, was a partial skeleton of a bowhead whale. This community is one allowed to subsistence hunt the great whales. I wanted badly to take the massive skull back to Texas but it was way too big for any airplane I might be in.

Toward the end of the brief Arctic summer, all volunteers were given a week of R & R. We could be transported to any place we each wanted on the refuge. Douglas and David went char fishing in the Brooks Range. I and another volunteer who liked to bird watch were flown by helicopter to a small peninsula projecting into the Beaufort Sea. There we camped for a few days watching the spectacular squadrons of hundreds of eider ducks pass hourly only a few feet over our heads. We hoped, sort of, for a glimpse of polar bear but none appeared.

At the end of our holiday, we were picked up by a single-engine float plane that had to taxi around small ice flows. On the flight to Barter Island, I rode beside the pilot—sort of stuffed in there. When we reached the island it was fogged in, so the pilot asked me to keep my eye on the beach edge below. I could see it because we were flying only 30 to 40 feet above the water surface. I was to visually guide him. I think they call this flying by the seat of your pants. All of a sudden the pilot shouted a curse and jerked the plane to the right . Our left wing hit an antenna, knocking it down, and the plane out of control.

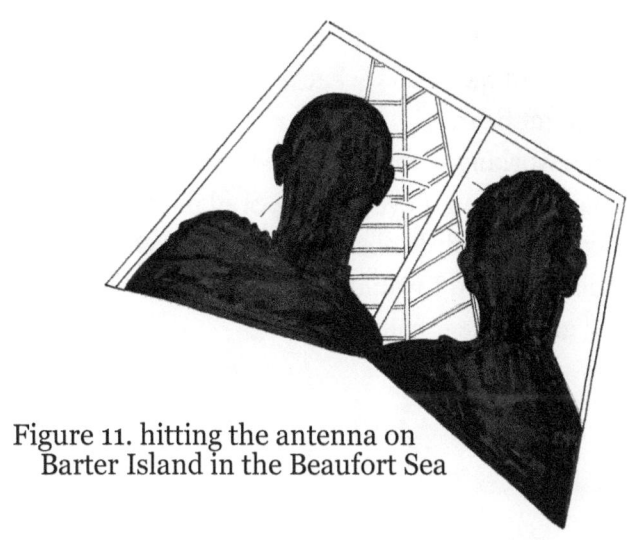

Figure 11. hitting the antenna on Barter Island in the Beaufort Sea

I gripped my door handle and squeezed my eyes shut, certain that this was it for me. It wasn't going to be a revolution in Guatemala, malaria in Peru, a pit viper in Ecuador, or a DC3 landing on lake ice that would pull me up short of a long life, but rather a float plane wreck in the Beaufort Sea. To this very day, I don't know how the pilot did it, but we landed float-side down on the bay. Everyone at the refuge headquarters came running down to see if we were all right, and we were. Fortunately, it was a small antenna that gave way easily without tearing off the wing. If I had had wealth, any at all, I would have gladly given it all to that bush pilot.

We worked another couple of weeks on our Aichilik plots before the summer's work was done. We were transported by helicopter back to Kaktovik, and that flight was to be my last adventure on ANWR. The Vietnam veteran pilot was one of those who enjoyed terrifying passengers by flying a laboring helicopter straight up into the sky and then straight down toward the ground before leveling out. He got what he wanted, a good chuckle, and I was terrified.

The Arctic is a different experience for the temperate zone naturalist; relatively few species but large populations of what is there— just the opposite of the tropics. It may well have been the most dangerous environment that I have experienced. Grizzly and polar bears are a genuine threat. Even with jaguars and venomous snakes, I never felt threatened by nature in Peru. The hair never stood up on my neck there like it did on the North Slope as we backed slowly away from a tundra plot and its resident bear.

It was enjoyable working with the US Fish and Wildlife Service professionals, and I gained much. I even coauthored the 1986 progress report with species accounts of birds observed in 1985 on the coastal plain of ANWR. But despite that, I would not, I think, choose to return. I really don't relish cold grizzly bear country.

Esmeraldas to Zamora Chinchipe

Ecuador called. Actually, Mark Robbins called. It was 1988 and he was collection manager of the extensive bird specimen holdings at the venerable Academy of Natural Sciences in Philadelphia. The neotropical ornithologist Robert Ridgely was a research associate there and author of *The Birds of Panama*. Ridgely and the Academy were working on a new project that would culminate in the 2001 publication of *The Birds of Ecuador*. I was going to get the opportunity to help again on the bird continent, and I couldn't imagine a better place. Ecuador sits astride the equator and ranges from the coast of the Pacific Ocean up to high Andean volcanoes and then down eastward into tropical rainforest.

My efforts were confined to boreal summer months (usually July and August) from 1988 through 1991. From San Angelo, via Dallas and Miami, I would fly into Mariscal Sucre Airport in the capital city of Quito, easily the most dangerous landings I have experienced in a commercial jet. The landing glide path was around an extinct volcano and near the slope of the smoking active volcano Pichincha. Looking out my right side window it appeared that our wingtip was only a few hundred yards from the mountain we were circling. Due to thinner air at the high Andean altitude, landings had to be performed at much faster than normal airspeed. Safely on the ground, we would hang out in Quito a few days (often within view of the magnificent volcano Cotopaxi) while Mark secured agreements of cooperation with the Museo Equatoriano and permits from government authorities.

Up until the 1960s or 1970s, US scientists collecting natural history specimens in Latin American countries were allowed to take back to the US most or even all of what was collected. That changed. Since at least the 1980s, most of those countries required scientists to leave in-country at their own university or national natural history collections, anywhere from one of each species up to fully half of all that was collected. So, part of the permit process is negotiating arrangements for disposition of specimens.

Typically, once permits were obtained we headed out in a rented bus with driver. Our initial direction was either north or south out of Quito for some distance before getting much out of the Andes. Mark always knew in general where he wanted to investigate.

In 1988, we were in the northern Province of Carchi along the Colombian border. During our weeks there, we had three field camps at varying elevations. The highest camp was just below páramo, the above-tree-line grassland of the northern Andes. It was freezing at night, and specimens are hard to prepare with blue fingers. But it was at that site I experienced another of those rare events in a naturalist's life, the sighting of an Andean condor. It was brief, but enough for a lasting memory. The huge vulture's more than ten foot wingspan is greater than that of any other Western Hemisphere land bird.

The other two camps were successively lower in elevation until we ended up in low foothills. A colleague scouted out the lowest proposed camp and reported back that it was beautiful with a perfect camp site. When we got there about ten days later, using around fifteen pack horses with enough men to work them, we found that the site was now a huge area of felled trees that we literally had to crawl over, a logistical nightmare. Mark decided to go ahead and work there, which turned out to be a good decision. The bird fauna in surrounding uncut forest was interesting.

Our team in Carchi was Mark Robbins, Skip Glenn, Tracy Pedersen, Douglas Wechsler, and me. Tracy is a Philadelphia area artist who can also prepare a good bird specimen. She is the primary illustrator of Princeton University Press's 2003 publication *Birds of the West Indies*. I later commissioned from her a painting, of any bird of her choice. Now hanging in my home is a prized Pedersen painting of scarlet-bellied mountain-tanager. Doug is the heart and soul of VIREO, visual resources in ornithology at the Philadelphia Academy of Sciences. Doug would erect a thigh-high rectangular tent, festooned inside with shrub and tree branches. He would then release our living netted birds in the tent to get photographs by inserting his camera lens through a port in the tent. It was an ingenious method of obtaining photos of species that otherwise would have taken months of unbelievable effort, if even possible, in the forest.

In 1989 we worked further south in the provinces of Loja, Morona Santiago, and Zamora Chinchipe. We headed back north in 1990 to the province of Esmeraldas before returning to Morona Santiago. Then for my final summer in Ecuador, we worked mostly along the Pacific coastal region in Manabí, Guayas, and Azuay. On various of these expeditions, the team was Robbins, Angelo Capparella, Francisco Sornoza, Robert Ridgely, Tracy Pedersen, Doug Wechsler and me.

The stories that anyone could tell from field expeditions are limitless, so I will try to contain myself with a reasonable selection from four summers in Ecuador. I'll tell about horses, dangerous mountain roads, airplanes and forest landing strips, deadly bushmaster vipers, a murder investigation, toe fleas, and a lost-at-sea brush with death.

My uncle, Jess Frost, grew up in Wyoming helping his father horse pack elk and bear hunters into the Grand Tetons. So during WWII, the army used Uncle Jess as a muleskinner in the Alps of northern Italy. I would have otherwise assumed that even the 1940s were too late for the use of pack animals, but I came to find out that WWII wasn't even close to the end of it.

Pack horses and mules are common in Ecuador. My impression is that much of their use is in dragging logs, probably mostly mahogany, out of road-less forests. These are work animals, not pets, and by the standards of most of us from the US, not treated kindly at all.

In Carchi we moved camp using a horse pack train. The mountain trails were narrow and steeply sloped. At one point I was on foot at the rear of the line and approached a commotion ahead. A packed horse had fallen about thirty feet off the trail and actually landed on its back in the branches of a fallen tree. Amazingly, it was not visibly injured, but had it not been stopped by the tree would have fallen hundreds of feet to its certain death. One of the wranglers scrambled down slope, machete-chopped the horse out of the tree, chopped a new route back up, led the terrified animal back to our trail, rearranged the pack, and rump-slapped it on down the trail to continue its labor. The event was appalling to us but seemingly unremarkable to the horse drivers.

In Morona Santiago in 1990, we flew to a remote airstrip in chopped up forest where a horse train packed us a few kilometers into the interior. We arranged for them to return for us in a couple of weeks. The day they returned, two problems were immediately apparent. It was going to rain buckets, and among the animals not only was there a mare in heat but there was an uncut (uncastrated) stallion. Mark packed the prepared bird specimens carefully and wrapped the specimen trunk in plastic before tying it to a horse back. Given the impending problems, Mark and the others of our team headed for the airstrip. Then it rained, and rained, and rained. Mark noticed water gathering inside the plastic covering the trunk. He cut the plastic with his knife so it would drain, saving the specimens.

I remained back at camp with sacks of pots and pans, two horse wranglers, a mare, and an insane sexually-aroused stallion. If you have not been around horses in that state of arousal, let me explain the problem. Simply, such stallions are uncontrollable. Why on Earth, the men brought him to our task I'll never know. One man was occupied desperately trying to keep him under control. The other man was trying to tie the utensil sacs to the mare and get her down the trail and away from the stallion. We finally headed out of camp, and I spent the next two hours on the trail picking up tossed pots and pans in a soaking rain all the way to the airstrip.

At the airstrip, we had to wait several hours for our ride. The most interesting diversion there in the blazing tropical heat was an incredible cattle transport operation. An old DC3 airplane was being loaded with cattle! A wooden ramp led to the door in the side of the plane. I believe that the cows were blindfolded before being led up into the plane. When full, it took off, barely clearing high trees near the end of the runway. I suppose there's no other way to move cattle to market in that isolated part of the world.

I'm sure that if your livelihood involves working horses, then every horse that eats food you bought and takes up space you maintain for it has to pay you in performance. If it did not live up to those requirements, you and I in the US might try to sell it. In back-country Ecuador, apparently, they might shoot it in the head and eat it. That's what happened near our low camp in Carchi in 1988. Packing back out of that camp among felled tree trunks, our horses included another of those young stallions. I don't remember there being a mare in heat, but the stallion would not cooperate with the owner/wrangler. Out of sight ahead of me on the trail, I heard a pistol shot, and then quickly came upon the dying horse. Another wrangler explained to me that the horse flesh would not be wasted.

My last horse story actually involves a burro. In 1991 in Manabí we needed to pack train to the top of a small mountain—Cerro San Sebastian in Machalilla National Park. Included in all modern expedition equipment is a portable liquid nitrogen tank. Not only do we collect bird skins and skeletons but also frozen tissue. The tissue is used primarily to extract DNA for phylogenetic (evolutionary relationship) studies. To freeze tissue in the field, one needs liquid nitrogen, which airlines are reluctant to take on board, even in a sealed tank. So the tank is brought empty and nitrogen must be purchased in the destination country. Therein lies the problem in some parts of the world, but not in Ecuador. We always could find liquid nitrogen in Quito—it's used extensively in cattle ranching country to freeze vials of bull semen to be transported unspoiled for artificial insemination. Ecuadorian ranchers know liquid nitrogen, but most everyone else in that country doesn't know how to safely deal with it.

There we were at the bottom of Cerro San Sebastian with a pack team of burros. You have to balance pack weight on either side of the animal's back. One of the burro packers, in the process, disastrously let the nitrogen tank slide up and partially over the animal's back. Some amount of liquid nitrogen spilled onto the burro's pack saddle and must have gotten to the poor animal's skin. At -321° F, not only is it dangerously freezing cold, it also fogs in the atmosphere. Picture a frightened burro, largely invisible in a cloud of fog, kicking and bucking up the mountain with equipment flying in all directions and burro wranglers in hot pursuit. The humanitarian concern was for the burro.

The concern for the expedition's future was in saving enough nitrogen to continue the effort on the mountain. Both concerns came to amicable conclusion.

On top of Cerro San Sebastian we encountered an odd parasite: the world's smallest flea, *Tunga penetrans*. South Americans call it nigua. In humans, this minuscule flea burrows into skin, usually around the margin of toenails. Our mountain-top camp was near a home with pigs running around the grounds, and we received the gift of niguas from the pigs. It was not a particularly frightening problem, but annoying. Not being as flexible as babies, we could not pull our own feet up to our face, so we had to work on each other's feet. Our ritual of nigua extraction occurred in late afternoons when in pairs we sat face to face and presented to each other a foot. Using sewing needles, we dug the niguas out of companion's toes. It is a peculiar form of intimacy.

Without question, my scariest road trip in Ecuador was in 1989 in Morona Santiago. Our destination was a remote rainforest site on the Peruvian border. It would require military aircraft to get us to the village/outpost of Santiago, and that required heavy-weight connections. A US military attaché in Quito, known to Ridgely or someone else high up, arranged with the Ecuadorian Air Force headquarters for the flight. I don't pretend to understand the political clout required for him to pull off such an agreement, but he did. The next problem was just getting to the airplane. That required travel by bus on the most dangerous road that I ever traveled in the Andes. Between Baños and Puyo lies the road known today as the Ruta de las Cascadas, Waterfall Route, in the Pastaza Canyon. I assume it has been improved, because in 1989 it was known popularly as the Matador de Turistas, the Tourist Killer.

The Ruta extends thirty-eight miles from Baños to Puyo, dropping about 3000 feet in elevation. With little to nothing in the way of guard rails, a near vertical drop fell hundreds of feet to the river torrent below. The day we rode down was rainy and the road was slick. For me it was a white knuckle trip all the way. We did arrive safely, if weak-kneed, at the air base where some additional negotiations occurred. I gathered that there were some folks in the local Ecuadorian military command that were understandably less than enthusiastic about giving an expensive airlift to a ragtag group of US bird collectors.

The next day we loaded into a twin-engined STOL (short takeoff and landing) aircraft. Our flight to Santiago was forty minutes low over the rainforest. No one was kidding about the landing, it was short. Since rainforest was at the end of the ridiculously abbreviated runway, our approach was near vertically downward, and then after a split second leveling out of the plane, the pilot stood on the brakes.

We remained at Santiago for two weeks. A most interesting experience there was meeting the indigenous Shuar people. They are famous for earlier harvesting and preparation (shrinking) of human heads. The Ecuadorean government has long forbid the practice, so there were no worries in that regard. We received permission to visit the nearby Shuar settlement. As was my experience in Alaska, meeting these indigenous people was somewhat awkward and strained. We took a quick walk through their village and left. I am sure that they dislike being stared at.

One morning, I met a Shuar woman on the trail. She was less than five feet tall, bare-chested, and nursing an infant held with one arm while gripping a machete with the other hand. Strapped to her back was about a four foot long caiman, recently killed. Caimans are alligator-like crocodilians from the American tropics. My impression was that she had only that morning machete-whacked the caiman and was hauling it to market in her village. Clearly, I was not in the presence of your average Texas lass.

Toward the end of our stay in Santiago, we longed to extend our time there. Not only were the birds good, and more investigation would be fruitful, but we would do nearly anything to delay heading back up the tourist killer.

The province of Esmeraldas is in northwestern Ecuador, on the Pacific coast, and butted up against Colombia. It contains a southern portion of the Chocó wet forest that extends down the Pacific coast from the Darién region of Panamá. Some of the greatest annual rainfalls in South America occur there. Our 1990 site was not in the absolute wettest part of the Chocó, but it was wet enough.

Placing your tent is an issue in rainforest; a decision that one hardly gives ten seconds to in Central Texas. With your folded tent in your arms, you wander in circles while looking upward into the canopy because your tent cannot be placed under a tree limb. Limbs in wet tropical forest are smothered in bromeliads and other epiphytic plants such that the limb can literally weigh 1000 pounds or more. In a violent thunderstorm one occasionally hears a weighty limb crashing to the ground. There would be no chance for survival if you were beneath it.

After days of eating noodles and tuna fish in our Esmeraldas camp, some of us craved fresh meat. So we sent Francisco Sornoza to a village to purchase chickens, which he did and brought back alive. For a few days we had tethered chickens in camp. One of our team was a vegetarian, a difficult life style on remote expeditions. We did our best to accommodate his needs, but following such a diet is not easy to do out there. I cannot now remember how it happened, but he did not know that one of those chickens would end up in the soup. When he lifted up the pot lid to inspect the day's intended repast, he was aghast to see a de-feathered but otherwise complete chicken head (eyes, comb, and beak) and two yellow feet bob to the broth's surface. Ecuadorians are sensible people. They waste nothing of what can be eaten.

Near our camp was a small cleared spot, perhaps a half acre in size. I shot a trogon in that clearing, and unfortunately only wounded the bird. It scurried down a hole in the ground. Killing these birds is not fun and carries with it a moral responsibility to make every specimen count, but I was reluctant to put my hand in there. Imagination conjured a wide array of possible threats down that hole.

The next day a native man came into our camp wanting to sell a dead bushmaster. *Lachesis muta*, the South American bushmaster, is the largest pit viper in the world with records approaching twelve feet. It is a mortally dangerous animal, and probably in most cases its bite is fatal. As it turns out, he found the snake in that same trogon clearing. Bushmasters reportedly are territorial in behavior, so I assumed, now with goose bumps, that the snake had been there when I was. We decided that the Museo Equatoriano in Quito would want the specimen so we promised to pay the man. Although quite dead, what he brought in was still frightening. The serpent was probably eight feet long and with a head as broad as my hand. We always knew venomous snakes are found there, but still....

One of my strangest experiences occurred in Azuay in 1991. We spent a week in a small town, camped in a vacant building. Our collecting site was a forested ridge above town. With permission from village authorities, we erected our work tent in the central square, where we attracted considerable attention as we prepped specimens. We surely were the best entertainment in town that week.

One night, an Indian from higher in the mountains arrived at our tent with an odd request. He wanted photographs taken of a dead man. It seems that a man from his village went missing for weeks and then was found dead, his body reduced to mostly skeleton. Someone in the village suspected foul play and wanted photographs of the body wounds sent to authorities. They wanted a real CSI-type investigation but perhaps earlier had been ignored or quickly dismissed by Ecuadorian law enforcement. For some reason that I don't recall, it was decided that I and my camera would go with him.

The next morning it seemed as if he and I hiked miles into the mountains to reach the village. There I found what looked like a child's wooden coffin on the front porch of a house on stilts. There was a quietly grieving family in that home. As a foreigner I was not comfortable in the home. But I squared my shoulders, prepared my camera, and they opened the coffin. The skeleton still had some dried ligaments and muscle on it and was folded to fit the child-sized container, but what stood out was a sizeable hole in the forehead of the skull. I dutifully took several photographs and said I would find authorities that might look at them. Of course, neither I, my companions, nor anyone else we questioned knew to whom those photos should go to. I still feel guilty that I never did anything official with the pictures. The consensus among us was consistent with thoughts of some others in the village; the man most likely fell to his death.

Taking those photos has so far been my sole experience in crime investigation, and it won't bother me for it to remain that way.

My last Ecuadorian adventure is of an incident that could well have ended my life and those of my companions. After the toe flea episode on Cerro San Sebastian, we travelled to the Peninsula de Santa Elena in the Pacific coastal province of Guayas. Our intent was to make a collection of birds from the marine waters of Ecuador. It was necessary to boat away from the coast line. Typically, many marine birds—shearwaters, storm petrels, boobies and so on—don't come close to land. So, we rented a wooden boat with outboard motor and driver. Including the owner/driver, we numbered about six in this open, cabinless craft. I would guess now, with memory dimmed by time, that it was about twelve to fifteen feet long. What I do remember clearly is that none of us geniuses thought to bring water or food or to insist that there be oars for safety. We were, after all, going to be out on the Pacific Ocean for only a few hours.

The motor quit. We were adrift in the Pacific for about thirty-six hours, nearly out of sight of land, drifting seaward. The owner's brother in his own similar boat found us out there—how, I do not know. My fondest memory of the whole sobering affair was the small baleen whale that surfaced almost touching our boat in the middle of the night. My worst memories are of the psychological trauma developing among some of my companions. Being lost at sea can be a test of composure. There was even a suggestion that we try to dismantle part of the boat to make paddles. But the bird collecting out there was good and important—at least we had that.

Our efforts over the four summers of my participation added much to our knowledge of the birds of Ecuador. Summarizing the success of our mission is, I think, is less enjoyable and informative than telling about some bird groups that I came to know.

As a general rule on most of this planet, where there are trees, there are woodpeckers.

Oddly, that isn't true for Australia, New Zealand or Madagascar where that bird family does not occur. Most tropical forests, like those in Ecuador, abound in woodpeckers. The largest of them in the western hemisphere are in the genus *Campephilus*; the smallest, the piculets, are in *Picumnus*. In North America, the big campephilene woodpeckers include two of the most celebrated extinct species: Ivory-billed

Woodpecker in the US and Cuba and Imperial Woodpecker in western Mexico. About nine additional campephilene species occur from the northern American tropics down to southern Chile.

I found a powerful woodpecker in the Andes.

Figure 12. size variation within the woodpecker family

As with every other case of my encounters with these huge woodpeckers, this one made itself known by the hammering sound as it de-barked a fallen tree. Apparently, most *Campephilus* woodpeckers forage on beetle larvae that live under bark of dead or dying trees. It takes a lot of forest to have enough old dead trees to support a viable population of them.

In dramatic contrast to the big campephilines are the tiny (three to four inches long) piculets. They number some twenty-eight currently recognized species in South America. Reportedly there is a high incidence of hybridization among species in the group so how many real biological species there are remains to be determined. They're an odd group among woodpeckers; they don't excavate their own nest holes but rather use existing woodpecker holes. They also lack the stiff tail feathers known by all students of woodpecker anatomy. There is something special about a part of our hemisphere where the largest and smallest woodpeckers coexist.

The American tropics would not be what they are without toucans. I usually heard them before I saw them; they get together in chattering groups. The bill is one of the largest in relation to body size of any bird, and is the primary source of their fame. You would think that its bright colors and outlandish size would be for sexual signaling. After all, research has clearly shown that large tails in pheasants, grackles, and wydahs are precisely for that purpose. But in toucans, both sexes have large, colorful bills so sexual selection is unlikely.

Probably, the size (not color) of the bill is related to foraging. A toucan can sit on a branch and harvest fruit in a large volume of foliage without ever using energy to move, and it can reach out to pick fruit from twigs too small to support its weight. It can also reach into nests of other species and consume eggs and nestlings. Whatever the evolutionary impetus for that enormous bill, they are among the most distinctive of tropical forest birds.

In my four summers in Ecuador, I prepared seven species in this family—the mostly smaller aracaris, toucanets, and mountain toucans. I also prepared a toucan barbet. Barbets are the closest relatives to toucans, and that relationship is most evident with the colorful toucan barbet.

Figure 13. toucans and their close relatives

For many naturalists who work in tropical woodlands, caged parrots are unappealing. It's not the birds' personalities so much as it's the cage. Wild parrots are for me the most quintessential free tropical forest birds. They fly like the wind, reminding me of mourning doves in the Texas hunting season. They're loud and brash. A flock comes wheeling in, screeching for all its worth, then near miraculously disappears as it settles into a green tree. They simply disappear—it's the green thing. If you sit for a while, you may see them again when they begin that famous three-appendage creeping around branches. Along with two feet that they use like hands, they have a bill with both upper and lower jaws hinged. They can grip branches strongly with that bill as well as with their feet.

I prepped ten species of parrots. The largest was one of the green amazons, the mealy parrot. The smallest were two parrotlets in the genus *Forpus*. All of them smelled like belched fruit. One can often tell a lot about a bird's diet by its odor. Flickers in Texas often smell like formic acid, which they acquire from eating ants.

The diversity of birds in American tropical forests knows few limits. Bill diversity ranges from the usual cone shape of seed-eaters to the scythebills whose outlandishly elongate, curved bills should seemingly get in the way of almost every life function required of a bird. And then there are the big-headed puffbirds that nest in holes in the ground and that often catch and dine on tarantulas. The ocellated antbird is not only striking in appearance, but also noted for being the largest antbird that follows army ant swarms.

Ecuador is an addictive place for the naturalist. I spent more time there than in any other country on the bird continent. There is no end to what you can find or to the adventures you can enjoy or suffer.

Figure 14. endemic tropical American bird families

Rio Essequibo

The 1994 field season on the bird continent was to be very different for me. My future wife was a team member, and the expedition was funded in part by a grant awarded to Robert Dowler and me by our own school, Angelo State University. It had been my habit to work for others who secured the funding and ran the show. As I stated, this was very different.

Our destination was Guyana, one of only two mainland Caribbean nations and an impressive natural setting equal to Peru or Ecuador. It sits on the northern coast of South America and is underlain by the Guiana Shield, an ancient Precambrian granitic formation. Today, 80% of the country remains forested and 70% of that is reputedly in pristine condition. When we were there, most of the logging was selective for greenheart, a valuable tree, rather than clear cut for everything. Greenheart has remarkably dense wood, resistant to marine borers, and long used for pilings in harbors. Many species of animals facing trouble in other countries (like jaguar, giant river otter, and harpy eagle) remain with healthy populations in Guyana.

Guyana was for 200 years a British colony that initially gained its independence in 1966. That British connection explains why English is the principal language, although we did hear a lot of Creole spoken in Georgetown. Curiously, in 1994, close to half of the people there had their origin in another British colony: India. Their ancestors were brought to Guyana as indentured workers by the British. About 30% of the overall population was black, descended from Caribbean slaves. The remaining 20% were mostly indigenous natives.

Although the people in that country would rather it were not so, its fame in much of the world comes from the terrible day in 1978 when 918 people (including more than 300 children) committed suicide in the absurd Jim Jones cult colony of Jonestown. Guyana does not deserve that historical wound.

Our team that summer consisted of Mark Engstrom and Burton Lim from the Royal Ontario Museum (ROM) in Toronto, Canada, and Robert Dowler, Ann Boyd, and me from ASU. Mark had earlier been on the faculty in biology at ASU and was a long-time associate. Funding came from ROM and ASU with Smithsonian Institution contact assistance in Georgetown. Our objectives were to continue the inventory of mammals and birds in Guyana, goals of world-wide interest given the country's status as a veritable nature preserve. We worked two sites. The first was the Iwokrama Forest in the Potaro Siparuni Region and then Tropenbos, a Dutch forest research station, in the Upper Demerara-Berbice Region. Guyana doesn't have states; it has official regions.

Iwokrama is a large forest preserve that is an institution of the government of Guyana and the British Commonwealth Secretariat. The Smithsonian funded plant and animal research there as part of their "Biological Diversity of the Guiana Shield" program.

We were to gather in the capital of Georgetown in early July. Mark and Burton were old hands in Guyana, so they made arrangements for us in the country. Bob, Ann, and I flew there via Miami and Trinidad.

Georgetown is an amazing first impression of a former British colony. White clapboard houses gleamed in the tropical sun. Sewage in open drainage canals stank in that same tropical sun. Walking the streets, we occasionally encountered Rastafarians, tall black men with their hair in dreadlocks. And although dreads were known in the US since the 1970s with the rise of reggae music, we West Texans had rarely seen that remarkable hair style. We had rooms in what was at best a half-star hotel with no air conditioning, but we didn't really expect that luxury. A nice little café on the main drag had cool beer, Demerara rum, and good meals.

Ann carried in her backpack a double-bagged five-pound sack of corn meal—a necessity for specimen preparation. In the field it's used to soak up fluids, allowing for a cleaner prepared specimen. Only in this case, the rough handling of luggage somewhere in our journey resulted in a torn sack, and a backpack with corn meal spread liberally throughout—on toothbrush and hair brush, in clothing, everywhere. In reality, it was to be a minor inconvenience compared to what was ahead.

Obtaining government permits in Guyana in those years involved dealing with a limited number of people who on any given day might or might not be available. You just had to be prepared for delays. On the whole, for birds and mammals, a scientific collecting permit and an export permit for each group of animals were required by the Guyanese natural resource agency, and import permits were required by Canada and the US. Furthermore, we could not bring into our countries (US or Canada) what might have been taken without permission of the origin country (Guyana). So a legal requirement then was to apply for an import permit from the US or Canadian federal wildlife agencies, and that permit issuance was dependent on showing a copy of your Guyanese collecting permit. On this particular trip, given delays, we were advised by the particular Guyana official to go on into the interior with scientific collecting permit issuance to come later. That's what we did, more than a little nervously.

Getting into the interior of Guyana in 1994 was not a simple thing. There was one "highway," if you want to call it that, and but one way to travel that highway. The route was the Georgetown-Lethem Road, a mud track in those days. The way was in a Bedford truck operated by a transport company owned by Eddie Singh. Singh is a surname for followers of Sikhism in India.

Bedford trucks, built in Great Britain, are legendary. I can offer testimony to their ruggedness. Summer is the rainy season in Guyana. The ruts in the road were in places two to three feet deep, but the Bedford charged through them, unless it was too steep upslope in which case they played out a cable from a winch mounted on the front bumper, tied onto a tree, and dragged the truck upslope. Mind you, this was the major interior highway. We were told a story that I hope is true. The windows of the Bedford can be shut and the truck can then be driven across a narrow stream deep enough to submerge the cabin with only the engine air intake and exhaust extending above water. I believe it.

Figure 15. Guyana team awaiting departure

We boarded the truck in the early evening at our Georgetown hotel, then rumbled to Eddie's home. The canvas-covered cargo bed was then loaded with boxes and twelve people. Mark, Burton, Bob, and I crawled into the back with the mostly indigenous native passengers, but Eddie would not let Ann ride back there. He insisted it would be too uncomfortable and inappropriate for her, so she rode in the cab. Off we went—Bob, Ann, and I happily ignorant of what lay ahead. Mark and Burton kept their mouths shut.

Our truck bounced on through the night with all us cargo passengers trying to doze and me with a man's foot in my face.

Sometime after midnight we stopped. Eddie announced that we could go no farther until morning light, so we were to make ourselves as comfortable as possible. Ann slept in the cab. Eddie also slept in the cab, mostly out of fear of mosquitoes carrying malaria. Another bout of malaria was likely to kill him. His workman slung a hammock under the truck; nobody had told me to bring a bed for this road trip. Bob, Mark, and Burton slept with the people in the back of the truck. I stood around awhile trying to reason out the sleep problem before I dug a hole in the sand under the rear axle (the only spot without an occupied hammock) and curled up in it. I actually slept and had a minimum of oil dripping in my hair.

After crawling out of my sand hole, I found the first order of business in the morning was an alteration to the truck. The drive shaft was re-configured to turn the mighty Bedford into a four-wheel drive juggernaut. And off we went. In the dry season, we would have followed the Georgetown-Lethem Road all the way to our destination—near Surama. But that wasn't going to work in the wet season. Apparently, a considerable stretch in the middle of the route was impassible, even for the Bedford. So after many miles of slipping and sliding through mud, we detoured to the Rio Essequibo where our equipment was off-loaded from the truck.

The Essequibo is the largest river in Guyana. Only a bit less than 700 miles long, it is the longest river between the Orinoco and Amazon, and drains much of the interior of Guyana.

Our expedition was loaded onto two large canoe-like boats powered by outboard motors. Native Indians handled the boats, one man steering at the rear and one man up front looking for boat-destroying limbs, boulders, and other obstacles. Bob, Mark, and Burton, together with other passengers, took off early in the first boat. Ann and I boarded the second one, with all our gear, maybe an hour later. So much gear was on our boat that it had only about five inches of clearance to spare before slipping below the water surface. I thought that was probably all right until I spotted the approaching rapids. The Essequibo is famous for the frequency of its rapids and falls. Fortunately, the boatmen were experienced and calm. Nonetheless, shooting rapids can be a white-knuckle experience. I wisely had not explained to Ann's mother what I was getting her daughter into.

After seven hours on the Essequibo, we landed at Martin's Island. I am challenged to describe the incongruity of Martin's Island—a thatched roof, open-sided, dirt-floored café on a small island in the river. In the distance one could see only tropical forest with no obvious settlement anywhere close. Pigs and chickens ran freely everywhere on the island. We gratefully downed pony cans of Polar beer, Polar, in tropical Guyana, on the Essequibo River, in the middle of a remote rainforest. To top it off, an old memory of Ecuador was revived when Bob picked up toe fleas, niguas, from the pigs. We pitched a tent on the island and spent the night.

The next day a boat transferred us from Martin's Island to a landing on the Essequibo shore where we met Ultimate Warrior, another Bedford, this one driven by Eddie's son Brian Singh. The drive from here to our camp site in the Iwokrama was drier and easier but not without its moments. Our driver named his Ultimate Warrior for a reason: he seemed to need to humiliate the road with high speed. On more than one occasion, Ann, who again rode in the cab, was petrified as the truck kept its high rate of speed over bridges consisting of little more than two tracks of wooden planks.

We arrived shaken but intact at an old saw mill about three miles from the Mikushi tribal village of Surama in the Region of Potaro-Siparuni. There is today a major Eco-Tourist Lodge at Surama, but nothing of the sort was there in 1994. We set up tents under the saw mill's thatched roof and got to work. On this expedition, all of us but me worked on mammals. I collected birds.

We hired a couple of Mikushis, Sherry to cook for us as well as Harold for general camp aide. One day Harold brought us a freshly-killed paca. Pacas are rodents, large ones, the size of a terrier dog. Not only was it attractive, red with white spots, but it was delicious. Without burger or fried chicken emporia out in the rainforest, one loosens up and learns to appreciate what there is to eat.

The place was a naturalist's paradise. One clue to minimal exploitation in an American tropical forest is the continued presence of cracids. The bird family Cracidae is in the order Galliformes, the land fowl or chicken-like birds. Quail, turkey, grouse, and pheasant are familiar North American members of the group. The cracids, however, include birds probably unfamiliar to most people: curassows, guans, piping-guans, and chachalacas. Many of them are quite arboreal. Where indigenous people have access to firearms, cracids tend to disappear. Just like turkeys and chickens, they're tasty.

Guyana has at least six species of cracids. I saw guans and curassows in the Iwokrama. About every other day, small flocks of these birds would cross my path. Seeing these magnificent fowl in such numbers was truly exciting.

For the first time in my then twenty-seven years of working in tropical forests, I saw a column of swarming army ants on the move—on a raid as they say. Many of the suboscine antbirds are behaviorally specialized to forage at army ant swarms. They don't actually eat the ants; rather they catch the insects, spiders and other frightened invertebrates that incautiously flee the ant horde. Other antbirds, although not particularly specialized in this behavior, take advantage and opportunistically forage at the ant swarms. My favorite among this group was the white-plumed antbird.

From our camp, we could hear in the distance a lek of screaming pihas. Pihas are uniformly gray colored cotingas, a little larger than American robins. They're not particularly impressive visually, but the ringing sound they make is one of the most unforgettable sounds in any forest anywhere. It goes sort of "Qawee-qweee-oh." But the volume is ear-splitting if you're below the lek trees. I suspect a big piha lek can be heard miles away.

A lek is a gathering of courting males. Typically, among many lekking birds, by employing group behavioral display and vocalization, females are attracted to visit and select a mate. Ann and I visited the neighborhood screaming piha lek. She was impressed and remembers that lek and the sounds to this day. You also may have heard pihas anywhere else in the world, in movie theaters, as tropical forest background sound in movies.

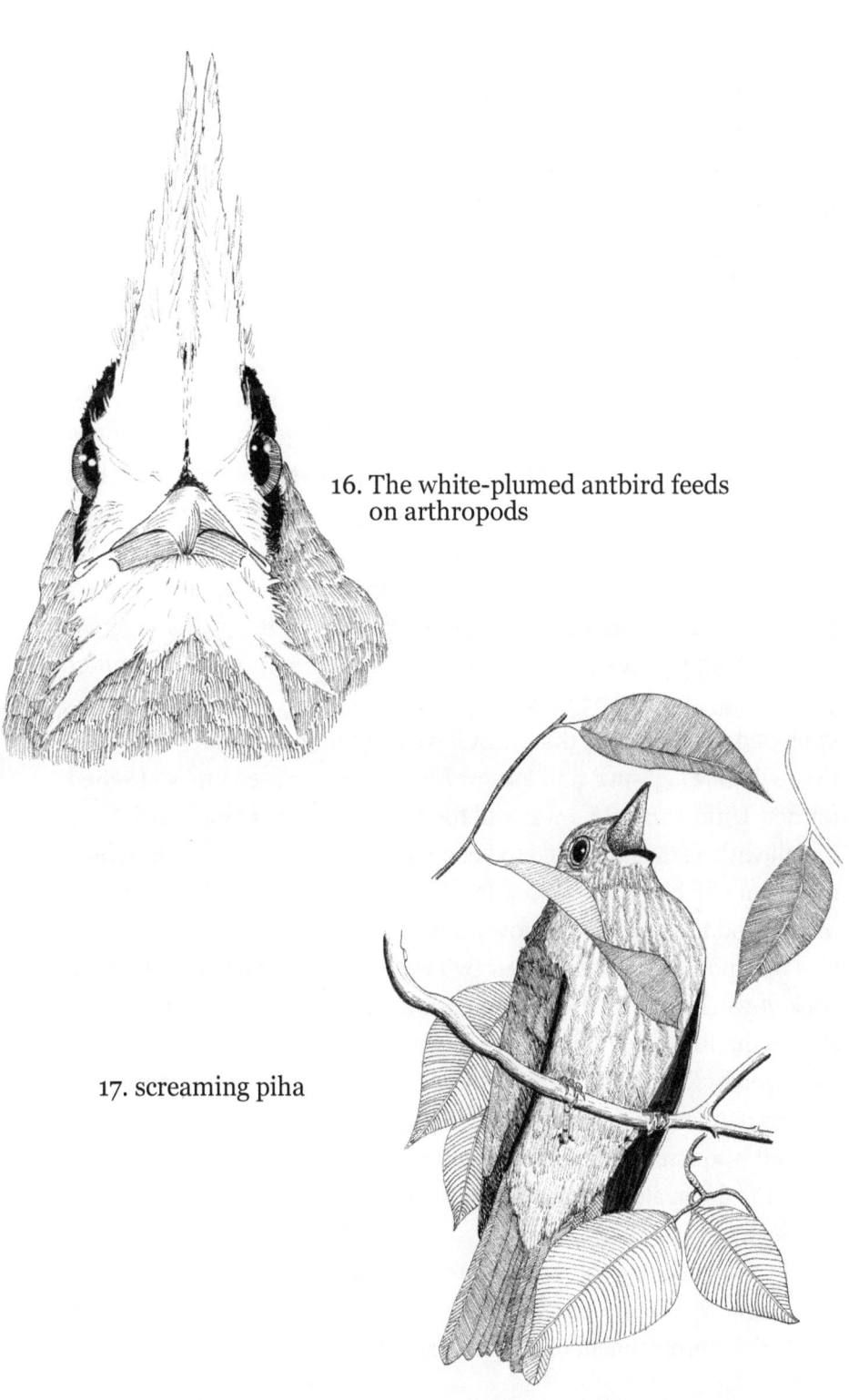

16. The white-plumed antbird feeds on arthropods

17. screaming piha

The best bird specimen taken in the Iwokrama was a male crimson fruitcrow. What sounded like a muffled dog bark "wuf" at the edge of camp turned out to be this big, bright red cotinga of northeastern South America. Although the species seems to be in reasonably good shape, population wise, mine was the first Guyana specimen in perhaps close to a century. Given its importance, I deposited the specimen in the LSU Museum of Natural Science, the largest collection of tropical American birds, where it will best contribute to knowledge of the species.

The mammal people used nets as much as I did, although their quarry were bats. One of their nets was placed over a fairly deep (chest level) pool, and required a dousing to remove bats from the net. The hottest camp talk one day was the discovery that a fairly sizable caiman (one of the crocodilian species) called that pool home. We bird people don't usually encounter such an issue with placement of bird nets, although I and others had been sponge-bathing at the pool's edge.

Walking back to camp from a bath one afternoon I encountered a large boa, almost certainly a *Boa constrictor imperator*. It was stretched across a road that must have been nine feet wide. It was my first wild boa sighting, and I knew Ann had never seen one so I yelled for her. Unfortunately, I was still too far from camp to be heard. I poked its tail with a stick and tried to keep it on the road while I yelled some more but still had no response from camp. I gave up when the boa tired of my prodding and turned toward me.

Upon completion of our two week's work at Surama, we traveled back downriver to the Upper Demerara-Berbice region on the opposite side of the Rio Essequibo. There we worked at a sand forest site managed by Tropenbos International, a Netherlands-based non-government organization dedicated to wise tropical forest management. A small wooden building had convenient work tables and floor area for throwing down sleeping bags. We were not familiar with such luxury in the field.

Mammal and bird sampling continued as it had in the Iwokrama. What impressed me the most at this site were venomous fer-de-lances, amphisbaenians, tegus, primates and royal flycatchers.

My experiences in Central America, Peru and Ecuador had resulted in only two encounters with the dangerously venomous viper called fer-de-lance. I had always wondered why I was not seeing more of this supposedly common snake. Well, that viper drought ended at Tropenbos. They were everywhere. One was beside our building and several were found by the mammal trappers in the forest. I shot one on Mark's trap line. A close call for Bob came when he reached to pick up a Sherman live trap only to see at the last safe moment a coiled fer-de-lance beside the trap. In fact, Bob seemed to be a magnet for this snake. Some of us seriously considered shunning him for our own welfare. We brought polyvalent antivenin with us, but were nervous about the safety of administering it, even in the event of a bite.

One of the joys of being a naturalist is the opportunity to experience oddities. Get a group of naturalists around a campfire or a pitcher of beer, and sooner or later a competition erupts into one-upmanship. Who has seen the oddest of the odd in the natural world? All of which brings me to amphisbaenians or worm lizards. They're reptiles and more specifically squamates, putting them close to lizards and snakes. One occurs in southern Florida. Otherwise you need to look for them in tropical America or Africa.

They're elongate like a snake or worm. One Mexican species has retained one pair of limbs; all the other amphisbaenians are completely limbless. They have no external ears and the eyes are imbedded deeply in the body surface.

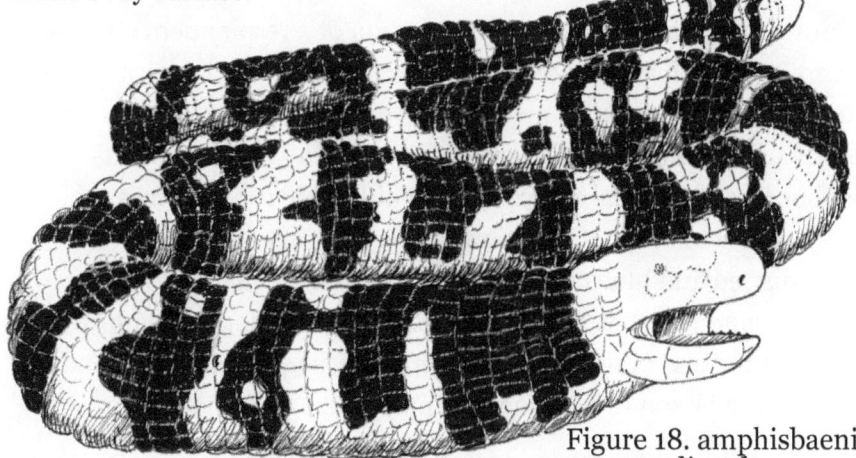

Figure 18. amphisbaenian or worm lizard

Since they live almost their entire lives in burrows they have dug, eyes are useless, but clearly they are descended from ancestors with uses for eyes.

Headed back to camp one day, I spotted a strangely black-and-white patterned snake-like reptile in the road. It seemed moribund, almost lifeless, and I picked it up. Back in camp Mark was examining it when it bit the thunder out of him. It was for all of us our first encounter with an amphisbaenian, and I've never seen another.

At my home in Texas, the most familiar lizards are the eight to nine inch long whiptails or racerunners with long tails and famed for their running speed. The reptile family is Teiidae. In the American tropics you can find massive three foot long teiids. They're called tegus. They are impressive lizards reminding one of the monitors in Australia and other parts of the tropical Old World. At Tropenbos gold tegus had the habit of getting into mammal traps. When released, they would stumble off into the forest slamming into everything in their mad dash to escape.

For the novice naturalist in the American tropics, one of the great disappointments is the paucity of monkey sightings. They're furtive and can be incredibly difficult to observe. But probably I saw more monkeys at Tropenbos than anywhere else in my travels. Most of them were spider monkeys and capuchins.

In the study of biology, one quickly gets around to sexual selection as a mechanism for evolutionary change. For example among male birds, the garish almost neon coloration of a resplendent quetzal or the ridiculously long tails of peacocks and many other pheasants are products over many generations of females choosing for mates those males in each generation with ever slightly grander accoutrements. At Tropenbos, I made acquaintance with an old familiar but always stunning product of sexual selection, the royal flycatcher. When he shows it, his red and blue crown crest is a wonder of the bird world.

The sand forest at Tropenbos was our last biology adventure in Guyana, but the end of its political adventure was still down the road and was to be a lesson in bureaucracy, patience, and responsibility.

Figure 19. royal flycatcher

When after weeks of grinding effort you stumble, dog-tired, out of the wild, the hope is that all the permits will be ready and you can stop briefly for your last semi-cold Guyanese beer and get on that plane bound for Trinidad. That was a pipe dream.

The permits weren't ready, and inescapably our scheduled flight departures for home were imminent. Contact with Smithsonian personnel identified a man in Guyana who could act as my agent and see the trunk of bird specimens safely through the bureaucratic maze in Georgetown and onto an airplane in my absence. The specimens would arrive in Miami where permits would have to be presented and inspections done. In the US, the Fish and Wildlife Service maintains a tight watch on wildlife being imported and exported. So I either had to fly back to Miami and negotiate the safe passage of the specimens or I had to hire a broker to represent me in Miami. Since I had already begun the semester's teaching at ASU, the broker had to be my approach. That small detail cost me $1000. Grant money was unavailable for this little expense.

I received a telephone call from a federal agent in Miami one evening to find out if any specimens in the trunk were of species that also occur in the US. I thought not, but I was wrong. There were pauraques, a caprimulgid that is widespread from South Texas through the tropics. Those specimens were confiscated, but clearly to the wildlife inspection agent I was not trying to hide anything. Nonetheless the entire trunk was inspected. When I next looked inside, I found the contents in disarray.

The end of this trip contained lessons for the field biologist working outside the US. Be ready for anything, obey all regulations for both the source and destination countries, and be patient. Looking back on our expedition in Guyana, I actually rarely think of the problems I had to deal with at the end. Instead I remember the rapids on the Rio Essequibo, Martin's Island, guans, currasows, giant red cotingas, antbirds at an ant swarm, and frightening fer-de-lances. Good times.

Islas Encantadas

One bucket list item for most naturalists is a visit to the Islas Encantadas (Enchanted Islands), better known in English today as the Galapagos Islands. Located in the East Pacific Ocean, some 600 miles west of mainland Ecuador, these mostly arid volcanic rocks are famed for their intriguing natural history and contribution to Charles Darwin's investigations of the evolution of species.

The islands were discovered in 1535 and subsequently used as a base for English pirates preying on Spanish ships, as a source of tortoise flesh for food on long sea voyages, and as a base for whaling ships. Ecuador annexed the islands in 1832, and colonization began. Darwin, on the round-world voyage of HMS Beagle, spent a couple of weeks there in September of 1835. At that time he had not formed well his eventual ideas on evolution, but observations of the distributions among the islands of tortoises and mockingbird species contributed to his growing doubts of the stability of species. Over the next 170 years, more colonists arrived, a US Army Air Force base was established, and in 1959 the islands became an Ecuadorian national park.

There are eighteen main islands and a number of smaller islets. The largest is Isabela at about 1,800 square miles. It has six volcanoes, one of which I would partially (and exhaustingly) ascend. Santa Cruz has the islands' main town, Puerto Ayora, where you can actually access your bank checking account through an ATM—this on an oceanic island no less! Is there any more startling testimonial to the global pervasiveness of technology than finding an ATM on Galapagos? Santa Cruz also has the greatest human population in the islands and hosts the Charles Darwin Research Station.

The flora and fauna are legendary and are the islands' primary attraction. For a botanically-challenged naturalist like me, the most impressive plants are the prickly pear cactus trees, and that is no exaggeration. Some of them appeared to reach more than twenty feet high with heavy wooden trunks. Among the more famous endemic animals are the world's only algae eating marine iguanas, massive tortoises, flightless cormorants with small wings too tiny by two-thirds to support them in flight, and the thirteen species of Darwin's finches with species-variable beak dimensions shaped ultimately by the islands' climate. Other fascinating animals there are colorful land iguanas, Galapagos penguins (the only penguin living naturally far enough north to encounter the equator), and four species of mockingbirds (each confined to different islands).

Figure 20. Galapagos mockingbird

My opportunity to experience these famous islands came in the year 2000 as a result of the work of my colleague Robert Dowler. Bob had been investigating the native rodent fauna there since 1995 and was becoming a fixture in the scientific community associated with the Charles Darwin Research Station. At 600 miles from coastal South America, rodents there are the only land mammals so far as is known to have ever dispersed naturally from the mainland to the Galapagos. There, adaptive radiation led to evolution of different species, often unique to different islands. The islands' native rodent fauna has been decimated by centuries of human introduction of non-native Eurasian rats, mice, and domestic cats. Bob rediscovered two species of endemic rice rats: *Nesoryzomes fernandinae* on Fernandina Island and *Nesoryzomes swarthi* on Santiago Island. He continued to explore the islands for native rodents through 2008, totaling seven trips by the end of his work. In our year 2000 expedition, the major goal was to search for rodents on Isabela and particularly at the rim of Wolf Volcano. As an ornithologist, I planned to do a behavioral study comparing the Galapagos Flycatcher to the local race of Vermilion Flycatcher. Our proposals were accepted and funded, and a dream was soon to become reality.

Our team consisted of Bob, Tom Lee, me, and Joel Brant. Joel had finished his Master's degree in biology at ASU only two weeks before our departure and was headed for more graduate work at Texas Tech University. Tom was also a graduate of ASU and then on the biology faculty at Abilene Christian University.

Bob, Joel, and I flew out of San Angelo on 12 June. We met Tom in Dallas. There began the unending occupation of traveling field biologists: managing heavy bags of field equipment (camping gear, traps, liquid nitrogen tanks). Airlines even then had restrictions on baggage weight. The four of us traded gear around in the airport terminal so that we each ended up with check-in of a sixty-five pound equipment bag and one personal bag. As it turned out, this was to be our easiest episode of baggage handling. We flew on to Ecuador via Miami, arriving in Quito near midnight.

Our first real problem reared its head early in Quito. Joel had sun-burned legs (from exposure back in Texas). They were in such bad shape with silver dollar-sized blisters that we feared his trip might be at an end without having really begun. So our first day in Quito involved contact with the Charles Darwin Foundation (whose influence got us reduced airline fare to the islands), arrangements for getting nitrogen, and hauling young Joel to the Clinica San Franciso. I went with him into the examining room to translate Spanish if need be. The doctor lanced the blisters, cut away dead skin, gave him prescriptions, and told him he was clear to go on the trip.

Early the next morning we taxied to the airport, flew to Guayaquil and went on to the Galapagos. We landed on the old US military runway on the island of Baltra. Within thirty minutes I saw a Darwin's finch! At almost the same instant I saw one of Darwin's famous mockingbirds, the Galapagos mockingbird. No doubt about it, I was well and truly in the Islas Encantadas.

After boating across the channel between Baltra and Santa Cruz and bussing through the highlands to Puerto Ayora, we arrived at the Charles Darwin Research Station (Estación Cientifica Charles Darwin). We spent most of two days here before heading to sea and Isabela Island. Doing biological investigation in the Galapagos Islands is, as I am sure anyone can understand, highly regulated by Ecuadorian concerns for the conservation health of the natural resources. Bob had frequent meetings with national park personnel and staff at the research station. Arrangements had to be formalized, i's dotted and t's crossed. No one does anything in these islands of world concern without clearing it with one or, usually, more authorities. We spent the better part of three days in Puerto Ayora before leaving for Isabela in the evening of 16 June. It gave me time to get to know some of the islands' famous wildlife.

I was delighted with the tameness of Galapagos finches around the station on Santa Cruz, although it didn't help a lot with identifying them. For the ground and cactus finches, plumage is remarkably alike among them. The most common one around the station was the small ground finch. I also saw a few of the large ground finch, medium ground finch, and cactus finch. I was struck by the shortness of their tails, the plump bodies, and seemingly weak flight.

Figure 21. large ground finch of the Galapagos Islands

Compared with many of the brilliantly-hued and spectacularly-adorned tropical birds I came to know on the mainland of South America, these little finches were pretty dull to the eye. But they nonetheless are the most memorable birds seen in all my years in the southern hemisphere. They are emblematic examples of speciation (evolution of species). From a probable single colonization event from the South American mainland, adaptive radiation has resulted in at least thirteen species in the Galapagos Islands today.

We saw various sea birds from every vantage. We even had brown pelicans, magnificent frigatebirds, great frigatebirds, lava gulls, and marine iguanas entertain us at our meals in the station open air cafeteria.

Most of our effort in Puerto Ayora was in provisioning for our time in the wild. An important figure in that effort was Don Ramos, our dependable camioneta (truck) driver, who saw that we got around to where we needed to make purchases. Not only is there an ATM in Puerto Ayora, but there is a supermarket. On a small island in the Pacific Ocean no less. We bought supplies for four or five people for ten days on Isabela, and spent $225, and we followed up with stops at a ferreteria (hardware store) and panaderia for pancitos (small loaves of bread) and puros (cigars). Our tents and sleeping pads we rented from station stores. In addition to these items, our gear included minimal spare clothing, scientific specimen preparation kits, cotton, cornmeal, bait (rolled oats and peanut butter), alcohol for specimen preservation, cooking utensils, karyotyping equipment (I'll explain below), and live traps (eighty small Sherman traps, forty large Shermans, and twenty-six Tomahawk traps). Then, of course, personal research items like binoculars and tape recorders for bird observation.

In modern scientific field study of organisms, there is more to gather than simply the cotton-stuffed specimen that you're familiar with in public museum displays. We often now obtain samples of chromosomes found in the nuclei of body cells. That effort requires specific equipment and solutions.

All this equipment had to be properly packed for safety and efficiency in the field, and so there we were again, engaged in the interminable repacking of gear. On our third evening in the Galapagos, Don Ramos picked us up in his camioneta at 8:45 pm and drove us to the docks to begin our adventure. Our gear and we were loaded into a panga (small outboard motor boat) and ferried to our at-sea home for the next ten days—the Beagle. The knowledgeable reader will remember that HMS Beagle was also the name of the ship that carried Darwin to the Galapagos in the 1830s.

In addition to us four Americans, the ship had a crew of three—the captain, a cook, and a first mate. The Beagle was about thirty-five feet long, and as far as I'm concerned it was not nearly big enough for some of the high seas we were about to encounter. The boat was to be our base of operations. If we could get back to it for the evening, then we ate and slept on it and ferried in a panga back to shore the next morning. If we were to be too far interior, then we camped ashore.

After a ten-hour voyage through very rough, gut-wrenching seas, we arrived the morning of 17 June at Isla Isabela. Now it was time to get some rodent survey field work done. The plan was to put Bob, Tom and the two cargadores out at Punta Albermarle, at the base of Volcan Wolf. They were going to attempt a climb to the rim top of the volcano where trapping might turn up some native rodents. Bob had experienced such success on two trips to a volcanic rim on Isla Fernandina in 1997. It took two panga trips to get the team and all their gear ashore on Punta Albermarle. As it turns out, they did not make it to the Wolf rim. Climbing Galapagos volcanos is unbelievably difficult. The surface is of crumbly, craggy, dangerous lava flow.

Joel and I on the Beagle sailed west looking for likely trapping locations. I wanted sites at some distance from Bob's and close to Volcan Ecuador, another of the six volcanoes on Isabela. We trapped at two locations on the northern coast, chosen because from the water we could see at least some vegetation within the tortured lava landscape. Working such landscapes can be treacherous, with the first problem just getting ashore. From the panga you jump onto wave-swept lava rocks while loaded down with traps. Unfortunately, at neither location did we catch anything but introduced black rats.

At our farthest point west, we spent a night in a quiet cove behind Punta Vicente Roca. There we were surrounded by high steep cliffs with many marine birds. It was serene and peaceful.

The bird life along the Isabela coast is truly stunning. Flocks of blue-footed and masked boobies were diving vertically into the ocean. On the rocky shore were Galapagos penguins and flightless cormorants. Some of the penguins I saw were actually at least several hundred yards north of the equator, the only such northern hemisphere penguin location in the world.

Seeing the cormorants was easily my high point for the birds on my trip. This is the largest cormorant in the world and is utterly incapable of flight with its comically small wings. Evolution of neoflightlessness (a flightless condition in an organism descended from a flighted ancestor) is a well-documented Pacific island phenomenon (notably among rails, but also known in parrots, pigeons, and others). Having successfully colonized remote islands without predatory land mammals or reptiles in the distant past, there may have been no subsequent need for the energy expense of flight. This is especially true for those birds that forage in a manner that does not require them to fly up into trees or travel distances. The Galapagos species of flightless cormorant is the only example of this phenomenon among cormorants. The entire range of this bird, numbering less than 2000 individuals, is on the islands of Fernandina and Isabela. The species' population is confined to a ridiculously narrow zone encompassing from only the immediate shore out seaward to no more than about 300 feet. Within that small area on only the two islands, the cormorants submerge to hunt for underwater marine life. No wonder these islands impressed Darwin.

After checking and pulling in our traps on the morning of 20 June, the Beagle took us back to Bob and Tom's position at Punto Albermarle. We expected them to be coming down off the volcano on this day. About two in the afternoon, Joel and I saw our team picking their way down the mountain. Their collection effort was a bust; there were no hoped-for native rodents to be had in four days of trapping on Volcan Wolf. Tom was ill with what seemed to be a cold infection, and both our companions looked like they had been to Hell and back. After an afternoon of specimen preparation, we all crashed in our Beagle bunks by nine p.m.

The next day was to be the most difficult physical exertion in my then fifty-four years of existence. All of us but Tom trekked back up the volcano. My stamina was no better than fair, but my legs were not really up to the exertion and began cramping. In all honesty, I probably should

Figure 22. flightless cormorant of the Galapagos Islands

not have attempted that climb; I gave myself a first class cardiac stress test. But clearly I survived it. After a near three-hour climb, we reached the height we wanted to work, rested a spell, and began a lateral trek following a dim trail on the volcano's flank.

On this trail that I really came to know the Galapagos tortoises.

Figure 23. huge Galapagos tortoise

It remains a wonder to me that these 500 pound (and heavier) turtles manage to thrive on the jumbled lava slopes. The effort required to move that bulk under the best of circumstances would seem doubly difficult and even prohibitive over lava boulders, but there they were. One odd little natural history note about them is that they seem to like the smell of peanut butter! I know this because they were attracted to our Sherman traps baited with that bread spread.

We spent four fascinating days on Volcan Wolf. The highlight was Tom's discovery of a cave-like lava tube that contained skeletal remains of a presumed extinct Galapagos giant rat (*Megaoryzomes sp.*). Don't get too excited about the adjective "giant." It was a very large rat only when compared to other rats. This rat's bones had been first found on Isabela in 1984, but ours were nonetheless significant finds. Giant rat remains are known from Santa Cruz in the Galapagos, but most regard that population as a different species (*M. curioi*). The Isabela form remains undescribed. It was pretty exciting stuff. And in honor of the find, Tom christened the lava tube "Pebble's Cave," honoring his wife's name.

Sadly, we found no living Galapagos rodents; only the introduced Black Rat went into our traps. Most of my field effort on the volcano was attempting to record observations of foraging behavior of Vermilion and Galapagos Flycatchers, although little of it was useful to my proposed study. The palo santo shrubbery allowed these birds too often to remain out of view.

Our trek back down off the volcano was energetically much better than the climb up but perhaps more treacherous. My right ankle was very sore and wrapped for stability. Lava boulder-jumping did not help. We made it to the beach in time for lunch on the Beagle before weighing anchor and moving west. Our objective was Punta Bravo on Isabela's west coast. Only an hour or two into the voyage, the seas became rougher and rougher, with waves reaching six feet and higher. The little Beagle was giving us a memorable if sickening ride. The high point of this unpleasant boat ride was the abundance of marine birds, especially storm-petrels. We eventually turned the corner at our old Punta Vicente Roca and found precious calm water in that now familiar protected bay.

The next morning we made it by boat to Punta Bravo by 9:30 a.m., went to shore, moved inland to make camp, set traps, explore the area, and crash early in our sleeping bags. A highlight of this day was a sighting of swallow-tailed gull, not only the world's only nocturnal gull but also reputedly the only fully nocturnal seabird of any family.

Trapping was a bust; we caught absolutely nothing. That was our last night on Isabela. Nine days on the island provided no evidence that any native Galapagos rodents persist there. That result was disappointing but useful to understanding the current status of the islands' land mammal fauna.

On the morning of 26 June, we began the fourteen hour Beagle journey back to home base in Academy Bay, Puerto Ayora, on Isla Santa Cruz. Four hours of the voyage were smooth, and ten hours were rough —really rough. Bob and Joel stayed in their bunks the entire journey, although on more than one occasion they were nearly thrown out of their beds and on to the floor by the violent rocking of the boat. The cook, Juan, somehow was able to prepare a meal for Tom, me, and a cargador: some delicious tuna caught off the back of our boat.

We spent three days in Puerto Ayora, busy at multiple tasks in preparation for another voyage to search for native rodents. I gathered some more flycatcher data. One major effort was cleaning, I mean serious cleaning, of traps. A strictly enforced regulation of the Galapagos Islands National Park is that anything used on one island must be thoroughly cleaned before taken to another island. Introductions of non-native organisms (such as seeds used to bait traps) to the islands has been a disaster. Traps had to be disassembled, scrubbed with soapy water and brush, dried, and reassembled—not a minor effort with 120 traps.

We loaded and boarded the Beagle late in the evening of 29 June. We slept there and departed early the next morning for Isla San Cristobal. San Cristobal is east of Santa Cruz and the southeast trade winds were stiff in our face. The seven hours of this voyage were the worst yet. Bob and Joel were green with motion sickness, and although the rest of us did not get ill, we tired mightily of being thrown around.

The saving grace was that sea birds were good: waved albatross, petrels, storm-petrels, boobies, and frigatebirds. I was thrilled, but I couldn't get Bob and Joel to raise their green faces to look at them.

After a difficult landing on the wave-washed lava shore, we trapped again, looking for San Cristobal rice rat, unseen in over 100 years. Our one night of effort came up empty. Bob had worked different locations on this island in previous years, but not the eastern shore. Probably this last effort has fully closed the door on any hope for finding the San Cristobal rice rat.

After our one night on San Cristobal, we left for the small island of Santa Fe. We made the four hour trip with the trade winds at our back making for a much gentler ride.

Our purpose for the brief visit to Santa Fe was to take two specimens of the still surviving endemic Santa Fe rice rat. This rodent had never been karyotyped. Karyotype is the term for the visual description of the chromosomes in an organism's cell nucleus. Comparison of chromosomes between species contributes to understanding evolutionary relationships among them. We set forty-two live traps and caught thirty-eight Santa Fe rice rats by the next morning. They were everywhere and unbelievably tame. I watched one get caught in a trap and then a second one tried to get into the same then closed trap.

Returning again to Santa Cruz and the scientific station, our brief boat ride was again blissfully gentle. The next two days involved meetings with staff and officials and a much-anticipated seafood meal at a restaurant down at the docks. We got seafood platters piled high with every manner of sea creature, including chiton. The fish and lobster were excellent. The chiton could have been chewed throughout a four-hour boat ride without making a dent in its hide. Chitons are primitive molluscans, and as a general rule of thumb mollusks are good to eat. Clearly, general rules of thumb have exceptions. Tom was sick all night after the chiton feast.

By 5 July, our expedition to the Islas Encantadas was drawing near to ending. We spent two nights trapping on Isla Baltra. You remember that Baltra has the aircraft runway on which we landed on our arrival. The Ecuadorian military gave us transportation assistance there. Baltra is a low, dry island with luxuriant desert shrubbery. Although trapping rodents was very successful, everything we caught was an introduced rat or mouse.

We now were within a week of departing the Enchanted Isles. Our final trapping effort was on our base island of Santa Cruz at the El Chato tortoise reserve, and again it was a bust.

The reader might expect us to consider our rodent survey effort a failure and hardly justifying the time and expense of our expedition. That is not the case though, as we contributed to a better understanding of what remains and does not remain of the islands' native land mammals that existed within the past 200 years. The picture was now at least partially clearer. Before you can state some species probably no longer exists, you have to investigate. My flycatcher study was, however, a regret. Too little quantitative data was obtained to analyze.

In partial payment for my supper during this Galapagos summer, I gave a bird specimen preparation workshop to twelve students who were staff and volunteers at Parque Nacional de Galapagos and the Estacion Cientifica Charles Darwin. I was assisted by Professor Betancourt of the station's museum and Marco Jacome, an ornithologist from the Ecuadorean mainland I had met ten years previously when I was working on expeditions for the Academy of Natural Sciences of Philadelphia. We prepped salvaged birds whose carcasses had been discovered by staff and visitors. I did a barn owl, dark-rumped petrel, *Oceanodroma* storm-petrel, and a red-footed booby. All in all, it was a productive day and it contributed to good public relations between us and the Ecuadorians.

On our last day in the islands, Bob received the bill for all our expenses: $9,289. We all heaved a big sigh of relief. We were not going to have to remain washing dishes to pay off our debt.

Our departure day, 13 July, turned out to be the goat rope we feared, loading and unloading 340 pounds of packed equipment. We loaded it all in the station's camioneta at 6:30 in the morning and hauled it to waiting busses in Puerto Ayora. We were instructed there to load it onto the top of an old bus, which we did only to find out that a new decision had just been made to not use the old bus. So we unloaded the 340 pounds from the old bus and loaded a new bus. We drove to the ferry landing between Santa Cruz and Baltra. At the ferry landing we unloaded the bus and stacked the equipment baggage onto the dock. The first ferry to arrive had its top cargo space full of bricks, with no room for our stuff. Then a second ferry arrived and was lashed to the outer side of the first ferry. We had to jump onto the top of the first ferry, traverse it, and jump across to the roof of the second ferry, with, remember, 340 pounds of gear. Our ferry crossed the channel to Baltra. We unloaded the ferry onto the dock and awaited the bus from the airport. On this new bus, we had to heft the baggage into a high door on the side. We unloaded the gear at the airport, hauled it into the departure gates and paid $168 in overweight.

Upon arrival in Quito, the van we earlier had arranged to haul our gear to a hotel did not arrive, so we used two taxis, one filled with baggage and one with us. None of this was a real surprise. Such rodeos are part of the expedition game.

On mainland Ecuador, we spent a couple of days at the Tandayapa Lodge on the west slope of the Andes. This was a pure fun trip to see highland birds. They were magnificent, highlighted by twelve species of hummingbirds including the unbelievable booted racquet-tail. The flight home had two small adventures left in our summer.

The first was the interminable problem with liquid nitrogen tanks. Bob had a small travel tank designed to keep tissue cold while actually empty of free liquid nitrogen. The lining of the tank absorbs nitrogen such that one can empty the tank but keep temperature sufficiently low for a few days. The problem is that it looks like a smoking bomb that we were trying to get on an airplane. Bob performed his best explaining and pleading routine, and the tank ended up coming with us. Our second issue involved an insufficient layover time in Miami. Tom, Joel, and I left Bob in Miami. He had to clear our rat specimens with wildlife customs officials, and there is no way to speed that up.

My Galapagos Islands summer was over. My fondest memories are of giant tortoises on the slope of a volcano, plump Darwin's finches, prickly-pear trees, and fearful boat trips between islands (perhaps the word fond doesn't really apply in this last case).

Bits and Pieces

Within the span of a reasonably long life, there should happen small events that add flavorful memory to the broader sweep of what has been usual. And it's not as if they were at the time momentous for anyone else in any other context, but you never forgot them.

For example, I showed a wild armadillo to the son of one of the modern United States' most famous antagonists. For more than four decades following World War II, a Cold War existed between communist countries, largely the Soviet Union, and the United States and other western democracies. It was marked by an arms race that included frightening nuclear weapon proliferation. In the late 1950s and early 1960s, the Premier of the Soviet Union was Nikita Krushchev. You might be old enough to remember that an American U-2 spy plane piloted by Francis Gary Powers was shot down over Russia in 1960. In that same year Premier Krushchev entertained the world with his famous shoe-banging incident at a UN debate in New York. Then in 1962 the Cuban Missile Crisis almost saw the US and the soviets come to what could have been horrific war.

In 1991, the year that saw the end of the Cold War, Krushchev's son Dr. Sergei Krushchev emigrated to, of all places, the United States. Today, he is a naturalized US citizen and is a Senior Fellow at the Watson Institute for International Studies at Brown University in Rhode Island. In 1999, he was a guest speaker at the E. J. Holland Symposium on American Values at ASU. Being the local naturalist on the symposium faculty committee, I always guided the speakers on an afternoon break tour of local landscapes. With Dr. Krushchev, we explored the headsprings of the South Concho River. On one of the dusty ranch roads I drove up beside a laconic nine-banded armadillo who kindly granted an extended, eager close inspection by the son of the former Cold War era premier of the Soviet Union.

Early on in the history of the Holland symposium, it was a common decision of mine to take our visiting speakers to the pecan-bur oak forest along the South Concho River. Being a dry zone Texan, forests with tall trees always beckon, always. Walking in the dappled shade of trees is ethereal for a dry-zoner. Among the most western nesting in Texas of eastern screech-owls, great crested flycatchers, Acadian flycatchers, yellow-throated vireos, and Carolina chickadees are in the forests bordering the Concho Rivers. Surely, anybody would be impressed. Then it was pointed out to me that our speakers commonly haled from states like Ohio, Pennsylvania, and Virginia—not Arizona, and they were not naturalists interested in chickadee distribution. The trees that impressed me were pretty much ho-hum to someone living in the eastern deciduous forest.

On the other hand, had I been alert, I would have noticed as we drove the guests from the campus to that forest, past miles of mesquite thorn brush and prickly pear cactus, that they were staring intently out the windows. They have tall trees in Pennsylvania; they don't have mesquite brush land. I am reminded of two points. Somebody in the world finds mesquite interesting, and you should ask your guests what they would like to see, not so much what would appeal to you. So for the remainder of my symposium nature guide career I showed them javelina trails through spiny brush, jackrabbits, and ranch headquarters. They were much more engaged.

One of my former graduate students grew up in Indiana. All biology graduate student advisors in Texas should at least once work with a student from the Midwest or East. I bring her up because she did not appreciate the active range management war on mesquite in the southern plains. She simply saw it as destroying trees and not as an activity useful to plains economy or ecology. Some historical context was called for.

When Lt. F. T. Bryan passed through the Concho Valley in 1849 in his surveying for a Pacific railroad route and when John Russell Bartlett led his US-Mexico boundary survey team through the same area in 1850, what they described of the flat plains was a mostly open savannah of grassland with widely scattered mesquite trees. The story of what happened to that grassland vegetation is now fairly well understood. Overgrazing, fire suppression, and drought combined to disturb the plains ecology, often resulting in the dense stands of brush we see today. Where mesquite is uncontrolled (meaning not burned, poisoned, cut, chained, or grubbed) today, the land is far less productive for ranching livestock. It may also reduce the amount of ground water available to streams and then ultimately us, although that point is debated among range ecologists.

The war on mesquite has consumed millions of dollars and promises to require investment of millions more with questionable chance of lasting success. None-the-less the battles, pasture by pasture, ensue. For my student with Indiana values about trees, the destruction of mesquites was a moral transgression. I tried to explain the rationale behind cutting mesquite and the local history of the plant. I don't know how successful I was. But the contrast between history of trees in Indiana and the dry plains of Texas would make for a good historical ecology lesson.

Another of my graduate students fell afoul of the law while doing nothing illegal. I cannot imagine how I could have saved him from the hassle. He had permission from every relevant authority, including the state parks administration, to study golden-fronted woodpeckers on San Angelo State Park by capturing and banding them for individual recognition. It was a technique critical to his study of their territorial behavior.

Capturing birds in mist nets requires early morning preparation so that net sets are in place before dawn breaks. So my student in the predawn darkness would park his car beside the state highway and carry his long aluminum conduit net poles through a walk-in gate into the park brush land study site. Undoubtedly, someone driving by saw him disappearing into brush with his poles. A report of this clearly nefarious activity made it to law enforcement, and apparently the thought occurred to someone that this was likely a marijuana-growing operation. What else could it be? Netting woodpeckers? The investigation cleared him of course, but it was a goofy distraction to what turned out to be an excellent study. Another remembrance of this student was his enthusiasm for tasting most everything he encountered in the field: ants, caterpillars, sundry other insects, whatever. He told me that ants have a citrus flavor.

It's bothersome to the naturalist what can live around you undetected and therefore unreported. It can be embarrassing. It's doubly embarrassing when an organism is first reported by a complete stranger to your region, one who just stopped by for a brief visit. That happened to me and my fellow local bird enthusiasts in July of 1994.

I received a telephone call from a woman in California. While visiting family in San Angelo she got into the field for a morning or two of birding. Upon completion of her trip, she called me questioning why varied bunting was not included on our bird checklist. I authoritatively explained that the species does not occur here. She then related where she had seen one—inside the San Angelo city limits no less. I warmly thanked her for the report, wondering at the time what she might have confused with this borderland bunting. But just on the off-hand chance of there being something interesting there, I called two ASU chemists who were also dedicated bird banders. They rushed out there before sunrise the next morning and set up mist nets. I got their telephone call not long after sunrise.

They indeed caught, banded, and photographed a bunting—the varied bunting. Our California visitor knew what she saw. Then there was a long pause over the phone. "We also caught something else." As it turns out, the bunting was to play second fiddle that morning. They caught an elf owl, another and even more spectacular borderland bird previously unknown from the Concho Valley.

The elf owl is one of the smallest owls in the world at only about five inches long. In 1994 it was known in Texas from much of the Trans-Pecos but only sparingly east of the Pecos and north of the Rio Grande. There were rare reports as far north as Sonora in Sutton County, sixty-five miles south of San Angelo. It was also known from the immediate Rio Grande around Laredo in South Texas.

Figure 24. elf owl, smallest of the world's owls

The find of an elf owl that morning begged an important question. Since bird records in the San Angelo region go back over 100 years, could this species have been overlooked all that time, or was this diminutive owl a new arrival? Is it part of the northward expansion of bird distributions coincident with global warming? Had the chemists captured the only elf owl in the Concho Valley—perhaps a lone visitor? Just how secretive can an owl be, even one this small? The following year, my dedicated search for this little owl turned up another one about twenty-five miles west of San Angelo in Irion County. Then from 2002 to 2007, a pair nested in a woodpecker hole in a utility pole, again in Irion County. It seems likely that they had been here for more than a couple of years anyway. I doubt we will ever have the answer to the question. And clearly, I should never dismiss a report from an out-of-town birder.

If a tiny owl can be secretive, then can also a beetle? So, if an organism lived below ground surface, but only in a soil-dung mixture and that only in the entrance runway to a prairie-dog burrow, would you call it secretive? I have seen such an organism, and I'm still not over my amazement, but I do forgive myself for missing this one.

In the 1970s, the largest remaining black-tailed prairie-dog town in Texas was close to San Angelo in northwestern Schleicher County. It was large enough to be home to thousands of the rodents. The full dog town ecology was present: golden eagles, ferruginous hawks, and prairie falcons in winter, American badgers, and western diamond-backed rattlesnakes were all thriving. I was contacted by a couple of entomologists at the Smithsonian Institution in Washington, D.C. who wanted to visit the location. They were on an extended trip searching in prairie-dog towns from Canada to Texas for some highly specialized beetles whose habitat I described above. Would I take them to the site? I did.

When we arrived they grabbed their gear that included specimen jars, shovels, and garden trowels and marched into the prairie-dog town. I held my breath as they reached bare-armed into burrows. Such burrows are notorious for housing rattlesnakes. With trowels they dug into the compacted prairie-dog runways leading downward into earth.

And there they found their quarry. I could have lived my entire life as a naturalist in this region without ever being aware of these peculiar scarab beetles. It certainly makes me wonder what else might secretively be out there. I also wonder how those beetles are doing today. Prairie-dogs in Texas are estimated to be down in number to only two percent of what existed in the late 19th century. And I would not be surprised if those numbers hold as well for the other animals dependent on the prairie-dogs.

I have no affection for caves. I find them dark, confining, and dangerous, although there are those who relish that environment, and cave-dwelling organisms do have fascinating adaptations. But since birds tend to ignore caves, I don't have to be ashamed of my aversion. Despite my usual avoidance of them, I do have two personal nightmarish cave stories.

In winter of 1973, I and a couple more fellow graduate students followed an ASU faculty biologist into two caves in Irion County. One of the caves, named Corn Grinder, is, as I remember, little more than a vertical twenty-foot deep sinkhole followed by a downward slanted passage of a few dozen yards. Our intent in entering Corn Grinder was to collect cave crickets, if present. We dropped a rope ladder down the sink and climbed down. The cave was small, dry and biologically uninteresting, so we soon returned to the surface. In leaving, I was the last one to climb up the rope ladder, and before I began the climb a message was relayed downward: "Uh gentlemen, we have a problem." Why we didn't see them when we were going down into the cave I don't know, but rattlesnakes were using crevices in the wall of the sinkhole as winter hibernacula. The passage wasn't much wider than a man's shoulder width, so there was nothing to do but go up right by them, inches away. I'm sure that despite the exertion of climbing a rope ladder, I held my breath the whole way. Nobody was bitten. Corn Grinder in my departing dust was the last time I saw, or ever will see, that cave.

I did enter another cave the following summer. The Devil's Sinkhole has been a protected (controlled access) state natural area since 1985 and is a National Natural Landmark. It is located on rural

Edwards Plateau limestone ranchland in Edwards County near Rocksprings. It scares me pea green and orchid, but despite that I allowed myself to be lowered into that awful hole in the ground.

The opening is a flush-with-the-ground hole that you cannot see until within maybe fifty to seventy-five feet of it. When you peered over the rim of the hole in those days you saw a dizzying 300+ foot free-fall drop to the bottom of a mountain of cave-in debris. Even more disturbing was the realization that when peering down into that terrible abyss you were standing on a rock lip projecting over the actual edge. Today, a safe platform for viewing has been built at the edge of the hole.

In 1974 I worked for a University of Texas natural area survey entitled *Devils Sinkhole Area—Headwaters of the Nueces River*. I was the ornithologist surveying the bird community. In the cave was a colony of cave swallows that nested near the entrance hole in the twilight zone. At the sinkhole portion of the project my job was minor, cave swallows. But I did enter the cave. Many of us on the team were not experienced cavers or familiar for any other reason with rappelling gear so a way was rigged to get us safely down that 300 foot drop. An Austin grotto (caving club) came to our need. A truck with an A frame backed up to the cave hole. A boatswain's chair was attached to a rope threaded through a pulley at the A frame's apex and the rope's other end (over 300 feet distant) was tied to another truck's bumper. The chair was pulled with a staff to the edge of the shaft, I was strapped into the chair and released to swing out over that empty-ness. Then the distant truck drove toward the hole lowering me into the Devil's lair. Coming back up was actually worse. When I was 300 feet up and they were using the staff to grab my chair, I was reaching desperately with my feet for the rim. It certainly makes for a good story but one that cannot be readily experienced today since entry to the cave is no longer permitted.

That graduate student of mine from Indiana who objected to mesquite cutting completed an interesting thesis research problem on neotropical migrant birds nesting in the Concho Valley. It was an analysis of how local woodland environments compared in usage by these migrants.

figure 25. into the maw of the Devil's Sinkhole

One of her study sites was in that pecan-bur oak forest near the head of the South Concho River that I mentioned above. She came into my office one day in the spring of 1996 to tell me she had seen a large, black hawk-like bird in the forest but had not seen it well enough to know what it was. I suggested that it likely was a common raven. She was certain it was not a raven and wanted me to come out to try to locate and identify it. So I did. She was right. It was not a raven. As with the discovery of that elf owl, I was flabbergasted. Her bird was a common black-hawk, a bird decidedly not common and considered threatened by the state of Texas.

We located a nest and later identified the nesting pair's favored hunting site on the South Concho River. This was an odd hawk that preferred to forage in streams where it could drop out of an overhanging tree into shallow water to grab water snakes, frogs, crayfish, and fish.

As a known nesting location, this pair's territory was a real outlier in Texas. It was the only nesting known east of the Pecos River and north of Laredo. We had brief glimpses of black-hawks near San Angelo in 1977 and 1978, but they were presumed to be visitors; we were probably wrong about that. Another common black-hawk nest was found five years later on Dove Creek only a few miles east of the South Concho site.

And then in May of 2006, I found another interesting hawk pair nesting a few miles north of San Angelo. I was part of a team assessing a possible submission of a proposal to get funding for a study of brush control. As the team's ornithologist, I would survey bird communities before and after brush clearing. Four or five of us in a pickup truck drove into the planned study site, and the mesquite was dense. You wouldn't try to ride a horse off-road through most of it. We stopped the pickup at a small pecan grove along a creek, stepped out of the truck, and were immediately assailed by high-pitched screaming of a hawk, a zone-tailed hawk. The nest was high above the truck in a pecan tree. I was stunned as I had been by the presence of the elf owl and the common black-hawk. At close to this same location, a pair of zone-tailed hawks had been found in the summer of 1953, over a half century before.

Figure 26. zone-tailed hawk (left) and common black-hawk

One of the men on the assessment team was clearly upset at finding this state of Texas threatened species. He would have much preferred had I not been present to identify the hawk. I understood his emotion even if I did not agree with it. He was honestly reflecting the fierce dislike of the Endangered Species Act, a common attitude in much of Texas and especially in ranching country.

This nest was far and away the most northerly reported in recent time for the species east of the Pecos River. Subsequent nervous whisperings have pointed to the likelihood of a small population of zone-tailed hawks in the rough canyons on the limestone divide between the Colorado River and North Concho River north of San Angelo in northern Tom Green County.

There is a greater point here than simply reports of populations of elf owl, common black-hawk, and zone-tailed hawk around San Angelo. It is that there remains a surprising amount of knowledge missing about the biota of Texas. We lack the opportunity, funding, personnel, and perhaps most importantly the permission of private land owners to adequately survey the geography of plant and animal life in Texas. All three of these species were found in the Concho Valley by serendipity. Who knows what would be found were we given the opportunity to look for it

Civilized Journeys of Exploration

We have all been reminded that sooner or later everyone slows down. I thought I never would. But when sleeping in a damp bag on the rain forest floor, arming yourself against grizzlies, and fighting insects that transmit life-threatening malaria or leishmaniasis finally begin to wear on your old bones, there are other options. You might as well embrace them, and some are as satisfying in their own ways.

The basic diversity of larger animal and plant life on earth was largely encountered and described for western science in the 18th and 19th centuries, and much of that effort was accomplished by Europeans, in particular British, French, and Germans. A university student studying zoology or botany today learns about the efforts of Joseph Banks, William Hooker, Alfred Russell Wallace, Charles Darwin, Jean-Baptiste Lamarck, the Comte de Buffon, Georges Cuvier, and Alexander von Humboldt, among others. Their efforts formed the foundation of the modern sciences concerned with biological diversity. Nothing better reinforces that foundation for students than to take them to Europe where they can experience with their own senses the fruits of their predecessors' accomplishment. Robert Dowler and I did just that.

In 2008, we assembled at ASU a study abroad program for Europe and put it into action in 2009. It was so rewarding that we repeated the program in 2011 and 2013. My offering was an undergraduate course of instruction entitled *The History of Western Biological Sciences,* and Bob put together a program on *The Role of Museums in the History of Science.* We took students to London, Paris, Eichstätt and Berlin in Germany, and Brno in the Czech Republic.

I won't belabor you with day by day, accounts of experiences in our explorations into the European history of science, but rather I have chosen those most memorable to me for their inspiration and often humor or angst in escorting university students abroad.

Along the southeast margin of greater London is the borough of Bromley, and in that borough is the village of Downe. In 1842, a young British naturalist, Charles R. Darwin, recently returned from a five-year voyage around the world, brought his growing family from London to live in the village in a home called Down House. Darwin resided there for the forty years before his death in 1882. To us naturalists, Darwin is the towering figure in our science. Although the concept of an evolutionary history of life was more than a century old at Darwin's birth, he put the concept on such a firm footing with his 1859 publication *On the Origin of Species* that his name is the one forever most associated with biological evolution. One can legitimately debate whether Darwin's *Origin* or Isaac Newton's *Philosophiae Naturalis Principia Mathematica* is the most influential science book ever written.

Down House is well preserved and furnished as it would have been in Darwin's day. Bob and I were always cautious in our expectations of students' reaction to the relevance and import of Down House, but we were seldom disappointed, for most liked what they saw. Darwin had a walk constructed in the woodland at the back of his property. There he frequently strolled for a quiet, stimulating getaway from his office in the house. Who knows what revelations of significance to biology he conceived on his famous sand walk? Without the slightest prior admonishment, our students walked quietly and reflectively on that famous 150-year-old sand walk.

Thirty miles northwest of London, in county Hertfordshire, is the town of Tring known for the presence of an old Roman road. We took the students to Tring because there you can visit an astounding little museum—the Natural History Museum at Tring. It was originally the passion of the 2nd Lord Rothschild who was famed for his zebra-drawn buggy. To this day, the mounted museum specimens in antiquated display cases are an historical treasure. We actually made the thirty-mile train journey out of London not so much for the Rothschild collection as to see the national museum research specimen collection of birds and, perhaps surprisingly, a library.

You probably have to be into this stuff as much as I to appreciate being within touchable distance of scientific stuffed skins of Darwin's finches, prepared by Darwin and Robert Fitzroy, Captain of HMS Beagle when Darwin visited the Galapagos Islands in the 1830s. And then there are the pigeon specimens prepared by Darwin. They were part of his extensive investigation of artificial selection, that is, human induced change in organisms.

Down a steep stair we were allowed entry into the most remarkable zoology/ornithology library I will ever see. The early volumes were accumulated by Rothschild, and, it would seem, to hold every historically significant book published on birds as well as many other zoology topics. At my request, they brought down from the shelf for inspection an original of *Systema Naturae*, 10th edition, 1758 by Carolus Linnaeus. This legendary volume is the beginning of modern scientific naming and classification for all Earth's animals. For example, a bird common to Texas is the wild turkey whose scientific name is *Meleagris gallopavo* Linnaeus 1758. If you're not impressed, then maybe you had to be there, and I will be forever grateful that I was.

Often, students of zoology have to be cajoled into an interest in plants, even though they know that Earth's animal life is utterly dependent on plant photosynthesis. And then you take them to Royal Botanical Gardens, Kew. This World Heritage Site is the largest collection of living plant species in the world. There is even a grove of giant redwood trees from the Pacific west of North America. Its most impressive green house is the Palm House, built in 1848 and considered to be the world's largest iron and glass Victorian building. Superlatives can go on and on about this stunning garden. Our Texas students were, gratifyingly, impressed. Sometimes you just have to look at a thing in the right place under the right circumstances to acknowledge its worth and beauty.

We sped in a bullet train under the English Channel to Paris, once the largest city in the western world. You must defend hauling biology students on the public dime to Paris when the subject is the history of science. Sure, the Eiffel Tower, Arc de Triomphe, Napoleon's tomb, and the world's most famous art museum, the Louvre, are in this great city and need to be seen by anyone visiting, but they are not directly related to the history of biology. However, tucked away obscurely in the 5th arrondissment (municipal district), in a corner of the Jardin de Plantes (France's main garden) is the Galerie des Paléontologie et d'anatomie comparée (the Gallery of Paleontology and Comparative Anatomy). That gallery was our primary reason for being anywhere in France, and in three summers students never failed to realize its importance to our journey of exploration.

The first floor houses hundreds of skeletons mounted in life-like poses. As you walk into the massive room, the skeletons all seem to be an organized parade moving toward you. It is easily one of the most impressive skeletal displays in the world. One could spend a lifetime in this room exploring anatomical differences and similarities among Earth's vertebrates. Many of the preparations date back to the era of Saint-Hilaire and Cuvier in the 18th century. On every visit, I walk to the rear of the parade where aquatic vertebrate skeletons swim in an invisible sea. There near the whales is a Steller's sea cow. When still alive around Bering Sea islands, this was the largest ever species of sirenian (related to manatees) but was commercially hunted to extinction by the end of the 18th century. Very few naturalists ever saw a living one, and regrettably now none of us ever will.

The second floor has similar mounted skeletons, but these are fossils of ancient extinct animals. It is my favorite floor in the museum. So many stories of Earth's history are associated with these animals. For example, under glass is an articulated skeleton of *Lystrosaurus murrayi*, a proto-mammal that lived about 250 million years ago. Its fossils, found on Antarctica, Africa, and India, confirm the existence of the great southern continent Gondwana that broke up into those smaller continents on which its remains are found today. It was greatly satisfying to walk students around the hall while relating some of the meaningful stories behind these extinct beasts.

Our first trip to Paris was also our most romantic in that one of our students proposed to his future wife while looking over the city from high on the Eiffel Tower. That's about as good as you can get for a proposal of marriage. And when you think about it, it's also biology.

From Paris, we went to southern Germany. Before I relate why we went there, let me tell you about getting there while shepherding better than a dozen American college students. In the first place, many of these young people are set in their ways. They resist awakening earlier than late morning. They want to eat only what they are familiar with. One young woman searched out a location for chicken tenders or hamburgers for every evening meal: in London, Paris, Berlin. Nothing would do but chicken tenders. As any traveler to Europe will tell you, moving around that continent is done efficiently on trains, but it requires that you the passenger be equally as efficient. The trains don't wait. If published departure time is 10:15 am, the train pulls out of the station at 10:15 am. It doesn't matter if you get out of bed too late to put on your makeup or sleep off your beer fest the evening before, the train will leave at 10:15. You must adjust your normal behavior patterns to the schedule of the trains.

In Mannheim, Germany, we had fifteen minutes from arrival on one train to departure on another, and of course the departing train was at a different loading platform from our arriving train. On one such transfer, I grabbed a dawdling student by the arm and pulled her forcefully into the train as the doors slammed shut loudly within inches behind her. Actually, our students for the most part adjusted to the needs of the schedule.

In the mountainous southern German state of Bavaria is a limestone formation called the Solnhofen. The limestones were originally laid down as sediments in lagoons near the sea in Jurassic age about 150 million years ago. Animal remains in those sediments were so exquisitely preserved that the formation is known among paleontologists as a lagerstätten (mother lode). Most famously, the only specimens of one of the early birds, *Archaeopteryx lithographica,* are found in the Solhnofen. We settled into a quaint hotel in the German town of Eichstät, taking a few days to investigate the fossils of the Solnhofen. We visited amazing small museums and pounded with rock hammers in limestone quarries. None of us found an *Archaeopteryx* but the students enjoyed so much the quest that had we gotten them there earlier in their lives, we might have made paleontologists out of the lot of them.

In 2009 and 2011 we went on to Berlin to visit the Museum für Naturkunde or Humboldt Museum, the largest natural history museum in Germany and one of the most important in the world. Not only does it present the world's largest mounted dinosaur skeleton but it displays what unquestionably is the single most important non-hominid fossil in the world—the Berlin specimen of *Archaeopteryx*. It is rare the textbook of science for any school level in any country that does not figure a picture of this remarkable fossil, transitional between dinosaurs and birds. Coming from London and southern Germany, I had seen three of the then known ten specimens of this animal, and now I could gaze upon its ultimate exemplar. I found out later that some of the students were more excited about taking photos of me, mouth gaping, standing before this display, than they were of the famous specimen itself.

In 2013 we dropped Berlin from the itinerary and ended our journey instead in the Czech Republic. We went there to see the abode of one of the giants of biological science, a man who died virtually unknown among those who came to depend so much on his insight. Gregor Mendel was an Augustinian Friar in Brno in what then was the Moravian region of Austria-Hungary. He was good at mathematics and interested in biological inheritance. His experiments with garden peas from 1856 to 1863 led posthumously to biologists' understanding of the basic principles of genetics. Those basics today give the obscure friar his just due in the history of science and are known as Mendelian Genetics.

Our students had an unrelated but important experience on the grounds of Mendel's Abby in Brno. Our tour of the Mendelovo Museum was led by a young Italian woman who was a college student in Brno and who communicated with us in fluent English. It dawned on our Texas students that she had to be proficient in three languages: Italian, Czech, and English.

When they asked her, with admiration on their faces, about her impressive language ability she was somewhat taken aback. "Do not you know more than one language" she asked?

My summers with students in Europe were deeply satisfying. It wasn't the Amazon or Beaufort Sea coast, but it was the historical underpinning for my interests in the natural world. It was also an opportunity to introduce that history to young budding physicians and scientists. When a young person tells you that they don't care for history, likely it just hasn't been introduced to them in the right way. To better prepare yourself for an appreciation of Darwin's place, you need to walk into his office in Down House and look at the desk on which he wrote *The Origin of Species*.

Good Times

My life as a naturalist has been both exciting and rewarding, but I am not so naïve as to believe that most people, even if they're naturalists, would want my experiences. Living some of these events vicariously in these pages would be enough to satisfy (or even horrify) most, and I understand. Clearly, I got myself into some scrapes that were unplanned and unwelcomed at the time. But there are young people who crave adventure beyond their cell phones and televisions. They want to experience, as a scientist, the organic smell of a rainforest. They want to see a toucan tooting from a wet limb draped with epiphytes. What should I say now to them? In brief, I say "go for it."

Much of my life has involved working with young people during that time when they are in preparation for careers. Many of them come to college with a goal painted too broadly. They might want it to be biology, but beyond that they're unsure if a more specific desire is feasible. They feel that they should be practical minded. A career in the health field makes all kinds of sense—medicine perhaps? Could it be in zoology? "Like what you do Dr. Maxwell?" "Where are the job opportunities in biology?" they ask, reflecting a common parental concern. "If I wanted to be an ornithologist, where could I make a living for a family?" They're all good questions, but ones that I hardly ever asked of myself at their age, making me perhaps one of the poorest career advisors in all of higher education.

I knew pretty much from the age of 12 what I wanted to do with whatever lifetime was granted me. I wanted to study birds. It's all I ever wanted to do. When I was sitting in an American literature class, I thought about birds. I wanted to do well in a chemistry class, not necessarily because I needed the knowledge, but rather I needed the grade to advance my pursuit of a degree that would let me study birds. What I wanted to tell my young advisees was that fixing on a specific goal is a powerful force during the low points of academic life, and there are always plenty of those, like taking organic chemistry exams.

I pursued my goal without a concern for how I was going to survive in the economic world, and I had no right to expect that I would actually ever be paid to study birds. But I did reach my goal, and someone did pay me to do it and to teach others about it. So, being a terrible advisor, I tell students to focus on what they would most enjoy doing and just go for it, hell bent for leather. Forget the job concern. "You're young—you can always regroup to plan B if your dream doesn't pan out." For this reason, I'm pretty sure I was unpopular with parents who were trying to keep their freshmen's feet on the ground. Feet-on-the-ground wasn't worth two cents to me.

I took some chances and had some close calls. Much of what I have related to you in these pages I never told my parents. There wasn't any point in doing so. My family and friends would have worried even more the next time I flew away for a summer in another hemisphere and to locations without phones or little if any medical help and with tropical diseases and revolutions.

My career life has been like a dream. I fear that I will wake up some morning and realize that I really did not see a condor high in the Andes, contribute however modestly to *The Birds of Ecuador*, show an armadillo to the son of Nikita Krushchev, or survey breeding birds on the North Slope tundra of Alaska. It has been the glad life of a journeyman naturalist.

Good times.

www.ingramcontent.com/pod-product-compliance
Lightning Source LLC
Chambersburg PA
CBHW020941090426
42736CB00010B/1222